MW00846681

THE ALGEBRA TUTOR

Vol. I

Algebra I and Algebra II

P. J. Thomas Book Company
An Independent Publisher
Jackson, Mississippi

ISBN 0-9660970-0-9

Library of Congress Catalog Card Number: 97-91181

Printed in the United States of America

PREFACE

The purpose of this book is to teach the basic skills of algebra on such a level that the average, and below average, student can get the main scope of an algebra course. The work that has been put together can be of vast assistance in Algebra I and Algebra II.

The motivation to write this book came from talking and interacting with students in my various summer school classes. I was often told by students that if I had been their teacher, they would not have failed their algebra courses during their main school year. They liked the manner in which I explained and demonstrated at the chalkboard. I would make a joke of their statements and say "maybe I should write a book and share some of my ideas and techniques with other people". As more and more students requested comprehensive written material on algebra, the more the idea became implanted. This is where it all started. I made up my mind and got to it. Each problem addressed has tried to be illustrated clearly and accurately with each step meticulously explained. I certainly hope that this book will prove beneficial to needy students.

I would like to thank my wife for allowing me to use her dining room table and chairs for so many months, my children for keeping me supplied with sharpened pencils and blank paper neatly stacked, and my daughter for her fantastic artwork used on the cover. Most of all, I would like to thank the Rosbury's, Floyd and Pat, for the great job of diligently typing each page to be sent to the printing company for the production of Vol. I. In addition, to Vol. I, now I have added Vol. II. All my thanks go to James Washington and his typesetters - LaGina Fisher, Jackie Williams, Cynthia Thompson, Emon Thompson, and Kamiliah McKinnon - of Sir Speedy Printing Center in Jackson, Mississippi.

ABOUT THE AUTHOR

WILLIE L. THOMAS

Teacher of Mathematics, Jackson Public Schools,

Jackson, Mississippi.

Mr. Thomas has twenty-one years of teaching experience and eighteen years of tutoring experience. Mr. Thomas received his Master's Degree in Mathematics, Elementary and Early Childhood Education from Jackson State University in 1975. He has participated in a National Teachers Corps Project through Jackson State University and has taught as a temporary instructor at Tougaloo College. His teaching experience ranges from second grade to freshman in college.

TABLE OF CONTENTS

ORDER OF OPERATIONS

1. Evaluate all powers including parentheses first.

2. After taking care of parentheses, work from left to right.

3. Do all multiplications and divisions before additions and subtractions.

4. Do all additions and subtractions from left to right.

INTRODUCTION TO THE ALGEBRA TUTOR

In order to understand and do well in algebra, one should learn the language and be familiar with the basic terms that will be used throughout the course. The purpose of this book is to give basic and clearer lessons on some of the most essential topics of algebra. The author's sincere goal is to write this book as if he is a personal tutor and give the feeling of someone sitting right there with the student to coax him or her into performing certain tasks.

Let's start with the language of algebra. We use various symbols in algebra and in order to interpret the language, one must be able to identify these symbols.

Example I:

$$12 - 4, \quad 32 \div 4, \quad 2 + 2 + 2 + 2, \quad 38 - 30, \quad 4 \times 2, \text{ or } 4 \cdot 2$$

[• means x] When computed out, all have the value of 8. We must remember that a numeral, or numerical expression, is another name for a number. The number is the value of the expression. We use the equality symbol, -, to show that two expressions name the same number. You write $4 \cdot 2 - 8$ and say "four times two equals (or is equal to) eight" and "8" is about the simplest name we can write for eight.

Example 2:

In working problems that have more than one operation involved. We must be able to identify with our grouping symbols, such as parentheses, brackets, and fraction bars.

GROUPING SYMBOLS

Parentheses

$5(4 + 3)$

Brackets

$5[4+3]$

Fraction Bar

$\dfrac{4 + 12}{8}$

Simplify $(23 - 6) + 5$

Solution: The parentheses ( ) tells us that "6" is to be subtracted from "23" and then, we are to add "5".

*Thus: $(23 - 6) + 5 =$
$17 + 5 = 22$ **answer**

*Note: $(23 - 6)$ and "17", both name the same number.

A **grouping symbol,** such as a pair of parentheses ( ) or brackets [ ] may be used to enclose an expression. We do not need to use multiplication symbols when using grouping symbols.

Thus: $5 \cdot (4 + 3) = 5(4 + 3)$ and $5 \cdot 4 = 5(4)$, which is the same as $(5)(4)$.

Let's look at $\frac{4 + 12}{8}$ the bar is a grouping symbol as well as a division symbol.

Thus: $\frac{4+12}{8} = \frac{16}{8} = 16 \div 8 = 2$ **answer**

Example 3:

Simplify $8 + [3 + (9 - 4)]$

Remember our <u>order of operations</u> tells us to start simplifying from the intermost grouping symbol and work our way out.

Thus: $8 + [3 + (9 - 4)] = 8 + [3 + 5] = 8 + 8 = 16$ **answer**

Let's Practice #1

Simplify each expression.

1. $3 + (5 - 2) =$

2. $(5 \times 2) + 3 =$

3. $7 \cdot (4 + 1) =$

4. $(16 - 5) \times 4 =$

5. $(12 \div 4)\, 6 =$

6. $18 - [3 \div (4 - 1)] =$

7. $\dfrac{70 - 7}{18 + 3} + 2 =$

8. $2 + \dfrac{40 + 12}{32 - 6} =$

9. $12 \times \dfrac{24 - 6}{9 - 3} =$

10. $32 \div \dfrac{9 + 3}{1 + 2} =$

11. $12 \times 6[4 + (16 - 6)] =$

12. $\dfrac{10 - [3(3 - 1)]}{7 + 2} =$

DEFINITIONS AND EXAMPLES OF ESSENTIAL TERMS

Through years of teaching experience, it has been observed that many students pass algebra and fail to develop a vocabulary with the basic terminology needed to build a strong background that will serve as a carryover from one course to another.

Variable -   A symbol used to represent two or more unknown numbers. Example: x, y, a, b, c, etc.

Constant -   A numerical expression without a variable.
               Example: 1, 5, 8, 12, 1/2, -1, -3/4, etc.

Real Number - Any number found or paired with a point on the number line.
               Example: 1, 3.11, 2/3, -1.0, etc.

The product of a Constant and Variable = An expression which gives the result of a constant being multiplied by a variable.
Example: "3 x n" is usually written "3n". This is the product of the constant "3" and the variable "n".

Numerical Coefficient - The non-variable factor in a term. In short, the number that sits in front of the variable or on the left side of the variable. (It is understood to be 1, if there is no other number showing.)

Example: In 3ab, it's 3. In x, it's 1.

Essential Tools To Work With In Algebra

It is very important that one must learn his or her sign rules in order to function well in a course of algebra. Listed below are the sign rules and examples of each. Practice saying the rules orally, practice writing them down and giving your own examples of each. Study and recite the rules everyday, until you know them as well as you know your home address. If you will do this each day for one week, it is guaranteed to improve your algebra skills.

Sign Rules for Addition Of Real Numbers

1.   (+)   + (+)   = +   Example: (+5) + (+3) = +8
2.   (-)   + (-)   = -   Example: (-5) + (-3) = -8
3.   (-)   + (+)   The first thing we do is subtract,
     or (+)   + (-)   Then we will keep the sign that is
                                  with the largest number.

3

You must repeat Rule #3 over and over because it is the one that gives students the most trouble.

Example: (1) $(-5) + (+3) = -2$

                                $(+5) + (-3) = +2$

            (3) $+25$

                                $\underline{-\ 7}$

                                $+18$

            (4) -14

                                $\underline{+\ 6}$

                                $-\ 8$

Sign Rule For Subtraction Of Real Numbers

Note: **There is only one rule.**

*Change the sign of the subtrahend (that part which is to be taken away), then proceed with the work just as you would do in addition. Example: <u>Subtract</u>

           (1) $+9$

                              $\underset{(+)}{-\ 7}$ (subtrahend)

                              $+16$ **answer**

           (2) $-\ 7$

                              $\underset{(+)}{-\ 6}$ (subtrahend)

                              $-\ 1$ **answer**

Sign Rule For Multiplication Of Real Numbers

1.	$(+)$ x $(+)$	$= +$	Example:	$(+3)$	x $(+5)$	$= +15$
2.	$(-)$ x $(-)$	$= +$	Example:	(-3)	x (-5)	$= +15$
3.	$(-)$ x $(+)$	$= -$	Example:	(-3)	x $(+5)$	$= -15$

Sign Rule For Division Of Real Numbers

1.	$(+) \div (+)$	$= +$	Example:	$(+8)$	$\div (+2)$	$= +4$
2.	$(-) \div (-)$	$= +$	Example:	(-8)	$\div (-2)$	$= +4$
3.	$(-) \div (+)$	$= -$	Example:	(-8)	$\div (+2)$	$= -4$

Working With Polynomials

Types of Polynomials

1.   Monomial - A single term which is either a numeral, a variable, or a product of a numeral and one or more variables.

                Example: $5a$, x, $2x$, $-3x^2$, etc.

2.	Binomial - A polynomial with two terms.
		Example: x + 2, a + 3, y - 2, etc.

3.	Trinomial - A polynomial with three terms.
		Example: $x^2 + 2x + 4$, $3x^2 - 6x + 9$, etc.

Anything larger than a trinomial is called a polynomial in that particular number of terms. For example $x^3 + 2x^2 - 2x + 5$ is a polynomial in four terms and that's all. The word "poly" (meaning much or many) used as a prefix on polynomials means "many terms".

Addition of Polynomials

Let's start with **monomials** first. Now, in order to add monomials, we must know something about similar terms and how to identify them. When we speak of similar terms, we are talking about monomials which have the same variable coefficient. For example, let's pick out the similar terms from the given set of terms: $5x$, $3xy$, $-2x^2$, $2x$, $8xy$, $3x^2$, $5x^2y$, and $-4xy$.

	Answer:	5x and 2x
	Answer:	$-2x^2$ and $3x^2$
	Answer:	3xy, 8xy, and -4xy
	Answer:	$5x^2y$

When adding any type of polynomial, we can only combine the similar terms. Let's add the given terms below.

$$\underline{5x} \; + \; \underline{3xy} + \underline{\underline{-2x^2}} + \underline{2x} \; + \; \underline{8xy} \; + \; \underline{\underline{3x^2}} \; + \; \underline{\underline{5x^2y}} \; + \; \underline{-4xy}$$

$7x + x^2 + 7xy + 5x^2y$ **answer**

	Example	1:	Simplify:	$2x + 3y + 3x + 7$
			Solution:	Look for similar terms and add.
			Answer:	$5x + 3y + 7$

	Example 2:		$3a^2b$
				$\underline{+2a^2b}$
				$5a^2b$

	Example 3:		$-12x^2y^2$
				$\underline{+7x^2y^2}$
				$-5x^2y^2$

Note:	Remember the sign rules for addition.

LET'S PRACTICE #2

Be sure to show all work just as shown in examples.

Reminder Note: Remember two terms are similar if they are exactly alike. This means the variable parts must be the same. The numerical coefficient can be different, this does not matter.

Example: $2x^2y$ and $-4x^2y$ are similar.

Pick out the similar terms:

1.   8ab, -4b, -9a, 7ab, 5ba, -3a

2. $2r^2s^2$, $-4r^2$, $3s^2$, $4r^2s$, $-6r^2s^2$, $-2r^2$, $3r^2s$, $2r^2$

Simplify:

3.   3a + (-3b) + (-2a) + b

4.   15x + 2y + 20x + (-3y)

5.   5p - 7q + 5q - 10p + q

6.   2a + 6b - 2a - 8b

Let's stop for a few minutes to talk about the distributive axiom which states a(b + c) = ab + ac. This means in order to move the parentheses, you must multiply everything inside the parentheses by what is in front of the parentheses.

Thus: (a)   3(a + b)   = 3(a + b)
                                  $= 3 \cdot a + 3 \cdot b$
                                  = 3a + 3b

(b)   -(+2) = -2

(c)   +(-3) = -3

Example 1:   Simplify 3(a + 2b) + (-1) (b + 2a)

(1)   Move the parentheses: $3 \cdot a + 3 \cdot 2b + -1 \cdot b + -1 \cdot 2a$
(2)   Find similar terms: <u>3a</u> + <u>6b</u> + <u>-b</u> + <u>-2a</u>
(3)   Add similar terms: a + 5b

Reminder Note: See sign rules for multiplication to move the parentheses properly.

Example 2: -5 (-3c + d) + 13[c + (-d)]

(1) Move the parentheses: +15c - 5d + 13 [c - d]   Note: (+) (-) = -

(2) Move the brackets: +15c - 5d + 13c - 13d

(3) Add similar terms: 28c - 18d

Note: Remember (-) +(-) = -, therefore   (-5d) + (-13d) = -18d

Now, LET'S PRACTICE SOME MORE. Be sure to show all work just as shown in the examples.

7. 4[h +(-m)] + (-2)(3h + 2m)
8. 5[a +(-b)] + 3(2a + b)
9. 3(-2a² - 3d) + [-(-3a² + 12d)]
10. 3[-2+ 7 (-3rp + 1)] + 5(12 - 6rp)

If you had trouble with problem 9 and 10 from LET'S PRACTICE SOME MORE", here are some sample look-a-likes.

Sample 1:   2(-3a² - 4d) + [-(-2a² + 13d)]

Remember the order of operations tells us to start working from the inside and work our way out. So, let's start by moving the parentheses first. Review the sign rules for multiplication.

2(-3a² - 4d) + [-(-2a² + 13d)]

-6a² - 8d + [+ 2a² - 13d]

Now let's move the brackets: -6a² - 8d + [+2a² - 13d]

Add similar terms: -6a² - 8d + 2a² - 13d

Answer: -4a² - 21d

Sample 2:   4[-3 + 6(-2rp + 2)] + 3(12 - 7rp)

First, move the parentheses:

4[-3 + 6(-2rp + 2)] + 3(12 - 7rp)

4[-3 - 12rp + 12] + 36 - 21rp

Second, move the brackets:

$$4[-3 - 12rp + 12] + 36 - 2lrp$$

$$\underline{-12} \quad - \underline{48rp} + \underline{48} + \underline{36} - 2lrp$$

Add similar terms:

$$-69rp + 72 \text{ or } 72 - 69rp$$

Remember -69rp + 72 is the most standard way to write the answer. This is with the constant coming last.

Addition of Binomials and Trinomials

When we add binomials and trinomials, we can use either of two methods; horizontal or vertical.

Example 1:   Horizontal Method:
Simplify $(x + 2) + (2x + 3)$
Let's start by moving the parentheses:

$$* \ (x + 2) + (2x + 3)$$
$$\underline{x} + \underline{2} + \underline{2x} + \underline{3})$$

*Note: If there is nothing in front of the parentheses, it is understood to have a positive sign in front.

Remember to add only the similar terms.

Answer: $3x + 5$

Example 2: Vertical Method

Simplify $(x - 5) + (3x + 8)$

Let's start by writing the expression to the right under the one that is to the left, Thus:
$$1x - 5$$
$$\underline{3x + 8}$$
$$4x + 3 \ *$$

*NOTE: The numerical coefficient of x is 1 when it is sitting alone.

P.S. Remember the sign rules for addition.

Example 3: Horizontal

Simplify: $(x^2 + 2x + 5) + (x^2 - 4x - 7)$

Move the parentheses: $(x^2 + 2x + 5) + (x^2 - 4x - 7)$

Add similar terms: $\underline{x^2} + \underline{2x} + \underline{5} + \underline{x^2} - \underline{4x} - \underline{7}$

Answer: $2x^2 - 2x - 2$

Example 4: Vertical

Simplify: $(x^2 - 3x + 7) + (x^2 + 2x - 8)$

$$
\begin{array}{r}
1x^2 - 3x + 7 \\
1x^2 + 2x - 8 \\
\hline
2x^2 - x - 1
\end{array}
$$

Answer:

Simplify: $(x^2 - x + 1) + (x + 3)$

$$
\begin{array}{r}
x^2 - x + 1 \\
x + 3 \\
\hline
x^2 + 0 + 4 \ *
\end{array}
$$

Answer:

* **NOTE:** 0 has no sign, you can write a "+" sign or a "-" sign, it doesn't matter, So, $x^2 + 4$ is the answer.

Let's Practice #3

Add the following:

1) $\begin{array}{r} 5x + 1 \\ \underline{x - 2} \end{array}$

2) $\begin{array}{r} 3x + 4 \\ \underline{x + 2} \end{array}$

3) $\begin{array}{r} 4x + y \\ \underline{x - 2y} \end{array}$

4) $\begin{array}{r} 2a + 3b \\ \underline{a - b} \end{array}$

5) $\begin{array}{r} 3a - 2b + 3 \\ \underline{2a + b - 2} \end{array}$

6) $\begin{array}{r} 3x + 2y + 5 \\ \underline{7x - y} \end{array}$

7) $\begin{array}{r} 5a^2 + 3 \\ \underline{3a^2 + 2a - 2} \end{array}$

8) $\begin{array}{r} x^2y^2 - z^2 \\ \underline{x^2y^2 + 2z^2} \end{array}$

9) $\begin{array}{r} x^2 + 2y^2 - z \\ \underline{x^2 - 2y^2 + z} \end{array}$

10) $2a^2 - 4ab + b^2$
    $\underline{2a^2 + 4ab + b^2}$

11) Simplify $(2a^2 + b - 4) + (a^2 + 2b + 1)$

12) $(3x + 2y - 1) + (x - 2y - 2)$

Subtraction of Polynomials

When we subtract polynomials, we must remember the sign rule for subtraction. That is to change the sign of the subtrahend and proceed as if we were doing addition. This means using all of the rules for addition.

Example:

<u>Subtract</u>		
$4x - 6$	$4x - 6$	
$\underline{3x - 2}$	$\underline{(-) 3x - (+)2}$	(Subtrahend)
Answer:	$x - 4$	

***NOTE:** Sign change shown in parentheses. Do not bother the original signs.

NOTE: When you change the sign of the subtrahend, this takes care of subtraction. Now, let's go back and add like we do in addition, using the addition rules.

This may seem strange to you, but it is the correct answer. Let's go below and check. Remember when you were in elementary school, we always checked our subtraction by using addition.

Thus: <u>Arithmetic</u>

12	(Minuend)
$\underline{-\ 5}$	(Subtrahend)
7	(Answer)

Now, to check we would add the answer back to the subtrahend. If the results gave us the minuend, we knew we were correct.

Let's check our answer: <u>Add:</u>

5	(Subtrahend)
$\underline{+7}$	(Answer)
12	(Minuend) IT CHECKS OUT

Now let's check the answer to the algebra.

Add:

$3x - 2$	(Subtrahend before the change was made)
$\underline{x - 4}$	Answer
$4x - 6$	(Minuend) IT CHECKS OUT

NOTE: Don't be lazy. Take the time to show the change of the subtrahend in a parentheses just as you see in each example.

Example 2:

Subtract:

$$3x^2 - 2x + 5$$
$$(-)x^2 + (-)3x + (-)7$$
$$\overline{2x^2 - 5x - 2}$$

Answer

To check <u>Add</u>:

$$x^2 + 3x + 7 \qquad \text{(Subtrahend)}$$
$$\underline{2x^2 - 5x - 2} \qquad \text{(Answer)}$$
$$3x^2 - 2x + 5 \qquad \text{(Minuend)}$$

The problem checks out correctly, therefore the answer is $2x^2 - 5x - 2$.

Let's Practice #4

We are going to use the same set of problems for subtraction as we did for addition. Our reason for this is to compare the answers for addition with the one's we get for subtraction.

<u>Subtract</u>:

1) $5x + 1$
   $\underline{x - 2}$

2) $3x + 4$
   $\underline{x + 2}$

3) $4x + y$
   $\underline{x - 2y}$

4) $2a + 3b$
   $\underline{a - b}$

5) $3a - 2b + 3$
   $\underline{2a + b - 2}$

6) $3x + 2y + 5$
   $\underline{7x - y}$

7) $5a^2 \quad + 3$
   $\underline{3a^2 + 2a - 2}$

8) $x^2y^2 - z^2$
   $\underline{x^2y^2 + 2z^2}$

9) $x^2 + 2y^2 - z$
   $\underline{x^2 - 2y^2 + z}$

10) $2a^2 - 4ab + b^2$
    $\underline{2a^2 + 4ab + b^2}$

Simplify:

11) $(2a^2 + b - 4) - (a^2 + 2b + 1)$

12) $(3x + 2y - 1) - (x - 2y - 2)$

Multiplication of Polynomials

NOTE: Remember to review the multiplication sign rules.

Before we get started, let's make note of some rules about exponents and variables.

** When we multiply **LIKE:** variables, we add their exponents.

Example: (a) $x^m \cdot x^n = x^{m+n}$, where "m" and "n" are real numbers.

(b) $x^1 \cdot x^1 = x^{1+1} = x^2$

NOTE: If you do not see any exponent on the variable, it is understood to have 1 as its exponent.

(c) $x^2 \cdot x^3 = x^{2+3} = x^5$

Now that we know what happens to the exponents, we can add them in our head and write the answer as such: $x^3 \cdot x^4 = x^7$. Now that we have made note on what happens to the exponents, let's take a look at an example that has a numerical coefficient involved.

Thus: (a) $2x^2 \cdot 3x^4 = 6x^6$

NOTE: We must multiply the numerical coefficients. Remember that is the number sitting in front of the variable.

(b) $(4a^2b)(2ab)(-3ab^2)$

$(4 \cdot 2 \cdot -3)(a^2 \cdot a \cdot a)(b \cdot b \cdot b^2)$

Answer: $-24a^4b^4$

** When we multiply **UNLIKE** variables, we **DO NOT** add their exponents. Examples:

(a) $x^m \cdot y^n = x^m y^n$, where "m" and "n" are real numbers.

(b) $x \cdot y = xy$

(c) $x^2 \cdot y^3 = x^2 y^3$

(d) $(x^2)(xy)(xy^2) = (x^2 \cdot x \cdot x)(y \cdot y^2) = x^4 y^3$

Let's Practice #5

MULTIPLICATION OF MONOMIALS

Simplify:

1) $a^4 \cdot a^6$

2) $b^2 \cdot b^5$

3) $(3x^2)(4x^3)(2x)$

4) $(-ab^2)(-a^3b)(-b^3)$

5) $(-4a^2)(-5a)$

6) $(-x^3y)(-2x^4y^3)$

---

7) $(1/3a^3b)(3ab^2)(a^2b)$ 8) $(a^n)(a^n)$ 9) $(a^{2n})(a^n)$

10) $x^n \cdot x^{3n+1} \cdot x^3$ 11) $y^{3n} \cdot y^{2n} \cdot y^n$ 12) $x^{n+1} \cdot x^n$

If you had trouble with any problem starting at #7 - 12, here are some sample look-alikes that may help you to master them.

Sample Look-A-Like for #7:

(a) $(1/2a^2b)(2a^2b)(ab)$ Since we have the fraction 1/2, let's write the rest of the expression over 1:

$$\frac{1}{2} \cdot \frac{a^2}{1} \cdot \frac{b}{1} \cdot \frac{2}{1} \cdot \frac{a^2}{1} \cdot \frac{b}{1} \cdot \frac{a}{1} \cdot \frac{b}{1}$$

Now let's group: $\left(\frac{1}{2} \cdot \frac{2}{1}\right)\left(\frac{a^2}{1} \cdot \frac{a^2}{1} \cdot \frac{a}{1}\right)\left(\frac{b}{1} \cdot \frac{b}{1} \cdot \frac{b}{1}\right)$

$$\left(\frac{2}{2}\right)\left(\frac{a^5}{1}\right)\left(\frac{b^3}{1}\right)$$

$$(1)(a^5)(b^3)$$

Answer: a^5b^3

Sample Look-A-Like for #8 and #9:

(b) $(b^n)(b^n)$

$$b^n \cdot b^n = b^{n+n} = b^{2n}$$

NOTE: Remember $n + n = 2n$

Sample Look-A-Like for #10 and #12:

(c) $x^{2n} \cdot x^{n+1} \cdot x^2$

NOTE: Remember we must add the exponents.

$$x^{2n + (n+1) + 2}$$

Move the parentheses: $x^{2n+n+1+2}$

Answer: x^{3n+3}

MULTIPLICATION OF BINOMIALS AND TRINOMIALS

Simplify: $(x + 2)(x + 3)$

There are several ways to multiply binomials and trinomials, but the two most used are the horizontal method and the vertical method.

Horizontal Method:

Using the distributive property:

$$(x + 2)(x + 3) = x(x + 3) + 2(x + 3)$$

Add similar terms: $\quad x^2 + \underline{3x} + \underline{2x} + 6$

Answer: $\quad x^2 + 5x + 6$

Vertical Method:

$(x + 2)(x + 3)$

Step 1	Step 2
$x + 2$	$x + 2$
$x + 3$	$x + 3$
$x^2 + 2x$	$x^2 + 2x$
	$\quad\quad 3x + 6$
	$x^2 + 5x + 6$

Answer: $\quad x^2 + 5x + 6$

Through years of teaching experience, it has been observed that students make less errors when they use the vertical method. You may use the one in which you feel most comfortable, but it is suggested that you use the vertical method most. Here's why:

(a)
$$x^2 + 2x + 1$$
$$\underline{x \quad - 3}$$
$$x^3 + 2x^2 + x \quad \longleftarrow \quad x(x^2 + 2x + 1)$$
$$\underline{\quad -3x^2 - 6x - 3} \quad \longleftarrow \quad -3(x^2 + 2x + 1)$$

Answer: $\quad x^3 - x^2 - 5x - 3$

(b)
$$x^2 - 2x + 3$$
$$\underline{x^2 + x - 2}$$
$$x^4 - 2x^3 + 3x^2 \quad \longleftarrow \quad x^2(x^2 - 2x + 3)$$
$$+ x^3 - 2x^2 + 3x \quad \longleftarrow \quad x(x^2 - 2x + 3)$$
$$\underline{\quad\quad -2x^2 + 4x - 6} \quad \longleftarrow \quad -2(x^2 - 2x + 3)$$

Answer: $\quad x^4 - x^3 - x^2 + 7x - 6$

NOTE: Remember the sign rules for multiplication and addition.

It is believed that the vertical method is more convenient to work with and thus minimizes the room for error.

Let's Practice #6:

Simplify:

1) $(x + 3)(x + 5)$ 2) $(x - 3)(x + 7)$

3) $(x - 1)(x + 3)$ 4) $(a - 4)(a - 3)$

5) $(a - 5)(a + 6)$ 6) $(x - 7)(x + 2)$

7) $(2x - 3)(3x + 1)$ 8) $(4a - 2)(5a + 6)$

9) $(3c - 4)(2c + 3)$ 10) $(x^2 + 3x - 1)(x + 2)$

11) $(a^2 - 2a + 1)(a - 3)$ 12) $(x^2 - 6x + 2)(x^2 + 3x - 1)$

Division of Polynomials

NOTE: Remember to review the division sign rules.

Before we get started, let's make note of some rules about exponents and variables.

When we divide **LIKE** variables, we subtract their exponents. Assume that no divisor signifies zero.

Example 1: $\dfrac{x^m}{x^n} = x^{m-n}$ where m and n are real numbers and m > n

Example 2: $\dfrac{x^1}{x^1} = x^{1-1} = x^0 = 1$ this is the same always, when m = n

NOTE: From expanded notations, we know that any number with an exponent of zero is equal to one (1).

Example 3: $\dfrac{x^3}{x^2} = x^{3-2} = x^1 = x$

NOTE: It is not necessary to write 1 as an exponent.

Example 4: $\dfrac{4x^4}{2x} = 2x^3$

Example 5: $\dfrac{8a^4b^6}{-2ab^2} = \dfrac{8}{-2} \cdot \dfrac{a^4}{a} \cdot \dfrac{b^6}{b^2}$

$= -4 \cdot a^3 \cdot b^4$

$= -4a^3b^4$

The following example demonstrates a special case of division, when the exponent m < n, in other words n > m.

$$\dfrac{x^m}{x^n} = \dfrac{1}{x^{n-m}}$$ where n and m are real numbers and n > m.

Example 6:

(a) $\dfrac{x^2}{x^3} = \dfrac{1}{x^{3-2}} = \dfrac{1}{x^1} = \dfrac{1}{x}$

(b) $\dfrac{x^5}{x^7} = \dfrac{1}{x^{7-5}} = \dfrac{1}{x^2}$

(c) $\dfrac{4x^3}{2x^4} = \dfrac{4}{2} \cdot \dfrac{1}{x^{4-3}}$

$= \dfrac{2}{1} \cdot \dfrac{1}{x}$

$= \dfrac{2}{x}$

When we divide **UNLIKE** variables, we do not subtract their exponents. Thus:

$\dfrac{15a^2b^2}{3ac^3} = \dfrac{15}{3} \cdot \dfrac{a^2}{a} \cdot \dfrac{b^2}{c^3}$

$= \dfrac{5}{1} \cdot \dfrac{a^{2-1}}{1} \cdot \dfrac{b^2}{c^3}$

$= \dfrac{5}{1} \cdot \dfrac{a}{1} \cdot \dfrac{b^2}{c^3}$

$= \dfrac{5ab^2}{c^3}$

NOTE: If the variables are unlike, we simply reduce the expression as low as possible and leave the exponents as they are.

<u>Negative Exponents</u>

Let's make some notes on negative exponents.

1. Negative exponents can be made positive by repositioning them from the numerator to the denominator and vice versa.

Thus: (a) $x^{-m} = \dfrac{x^{-m}}{1} = \dfrac{1}{x^m}$

(b) $x^{-4} = \dfrac{x^{-4}}{1} = \dfrac{1}{x^4}$

(c) $\dfrac{1}{x^{-m}} = \dfrac{x^m}{1} = x^m$

(d) $\dfrac{1}{x^{-2}} = \dfrac{x^2}{1} = x^2$

Let's Practice #7

Review the sign rules for division.

Simplify:

1) $\dfrac{8a^2}{2a}$

2) $\dfrac{3a^2}{a}$

3) $\dfrac{-9x^4}{3x}$

4) $\dfrac{x^9}{x^6}$

5) $\dfrac{-b^5}{b^8}$

6) $\dfrac{a^{13}b^{17}}{a^{10}b^{21}}$

7) $\dfrac{-16r^3t}{14rt^5}$

8) $\dfrac{-9xy}{-27xy^3}$

9) $\dfrac{12a^5b^8}{24a^2b^3}$

10) $\dfrac{-36u^5v^3}{27uv^5}$

11) $\dfrac{-16a^6b^2}{-24a^6b^5}$

12) $\dfrac{14x^2yz^3}{28xy^3z}$

Let's Practice #8

Now let's look at some practice with negative exponents. Write as a positive exponent.:

1) $2a^{-3}$

2) $\dfrac{a^2}{b^{-3}}$

3) $\dfrac{3x^4}{y^{-5}}$

4) $\dfrac{xy^{-3}}{5}$

Multiply and simplify the following:

Sample: $(a^{-2})(a^{-3}) = a^{-2} \cdot a^{-3} = a^{-2+-3} = a^{-5} = \dfrac{1}{a^5}$

NOTE: Review multiplication of monomials with positive exponents.

5) $(2a^{-3})(3a^{-4})$

6) $(5a^{-2}b^{-3})(3a^{-3}b^{-5})$

7) $(x^{-3}y^{-7})(x^{-1}y^{-1})$

8) $(a^{-1})(a^3)$

9) $(b^{-3})(b^4)$

10) $(x^{-1})(x^4y^{-3})$

11) $(x^2y)(x^{-3}y^{-5})$

12) $(3x^{-1}y^{-2})(2x^4y^{-5})$

If you had trouble with #12, here is a sample to look at:

$$(4x^{-3}y^{-1})(5x^7y^{-5}) = (4 \cdot 5)(x^{-3} \cdot x^7)(y^{-1} \cdot y^{-5})$$

$$= 20(x^{-3+7})(y^{-1+\ -5})$$

$$= 20(x^4y^{-6})$$

$$= \frac{20}{1} \cdot \frac{x^4}{1} \cdot \frac{1}{y^6}$$

Answer: $\dfrac{20x^4}{y^6}$

DIVIDING A POLYNOMIAL BY A MONOMIAL

Assume that no divisors are zero.

Examples:

(1) $\dfrac{4x+6}{2}$ $= \dfrac{4x}{2} + \dfrac{6}{2}$

$= \dfrac{4}{2} \cdot \dfrac{x}{1} + \dfrac{6}{2}$

$= 2 \cdot x + 3$

Answer: $= 2x + 3$

$$(2) \quad \frac{12y - 18y^2 - 6y^3}{6y} = \frac{12y}{6y} - \frac{18y^2}{6y} - \frac{6y^3}{6y}$$

$$= \frac{12}{6} \cdot \frac{y}{y} - \frac{18}{6} \cdot \frac{y^2}{y} - \frac{6}{6} \cdot \frac{y^3}{y}$$

$$= 2 \cdot 1 - 3 \cdot y - 1 \cdot y^2$$

Answer: $\qquad = 2 - 3y - y^2$

Example: $\quad (3) \quad \dfrac{16x^4y^2 - 48x^2y^4 - 24x^6y^2}{8x^2y^2}$

$$= \frac{16x^4y^2}{8x^2y^2} - \frac{48x^2y^4}{8x^2y^2} - \frac{24x^6y^2}{8x^2y^2}$$

$$= \frac{16}{8} \cdot \frac{x^4}{x^2} \cdot \frac{y^2}{y^2} - \frac{48}{8} \cdot \frac{x^2}{x^2} \cdot \frac{y^4}{y^2} - \frac{24}{8} \cdot \frac{x^6}{x^2} \cdot \frac{y^2}{y^2}$$

$$= 2 \cdot x^2 \cdot 1 - 6 \cdot 1 \cdot y^2 - 3 \cdot x^4 \cdot 1$$

Answer: $\qquad = 2x^2 - 6y^2 - 3x^4$

Example: $\quad (4) \quad \dfrac{24xy^3 - 16x^2y - 12x}{-4x}$

$$= \frac{24xy^3}{-4x} + \frac{-16x^2y}{-4x} + \frac{-12x}{-4x}$$

$$= \frac{24}{-4} \cdot \frac{x}{x} \cdot \frac{y^3}{1} + \frac{-16}{-4} \cdot \frac{x^2}{x} \cdot \frac{y}{1} + \frac{-12}{-4} \cdot \frac{x}{x}$$

$$= -6 \cdot 1 \cdot y^3 + 4 \cdot x \cdot y + 3 \cdot 1$$

Answer $\qquad = -6y^3 + 4xy + 3$

Let's Practice #9:

Simplify:

1) $\quad \dfrac{18a + 36}{9}$

2) $\quad \dfrac{24a^2 - 16a - 4}{-4}$

3) $\dfrac{6a^2 + 9a}{3a}$

4) $\dfrac{12a^3 - 4a^2}{4a}$

5) $\dfrac{x^2 + 2x}{x}$

6) $\dfrac{xy^3 + x^2y}{xy}$

7) $\dfrac{16a^3 - 8a^2 + 4a}{4a}$

8) $\dfrac{28x^2 - 21x^3 + 14x^4}{7x}$

9) $\dfrac{x^4y - 3x^3y^3 + xy^4}{xy}$

10) $\dfrac{3x^2y + 6x^3y^3 + 9xy^3}{3xy}$

11) $\dfrac{36x^4y^2 - 72x^3y + 24x^4}{-12x^2}$

12) $\dfrac{8x^4y^2 - 24x^2y^4 - 12x^6y^2}{4x^2y^2}$

LONG DIVISION OF POLYNOMIALS

When we do long division of polynomials, we go about it in the same manner as we do in regular arithmetic. Always start with the very first term of your divisor.   Example #1:

Step 1:

$$
\begin{array}{r}
4 \\
14\,\overline{)630} \\
\underline{56} \quad \longleftarrow \text{Subtract} \\
70
\end{array}
$$

Step 2:

$$
\begin{array}{r}
45 \\
14\,\overline{)630} \\
\underline{56} \\
70 \\
\underline{70} \\
0
\end{array}
$$

Example #2:

Thus:

Step 1:

$$
\begin{array}{r}
x \qquad \text{Multiply } x(x+2) \\
x+2\,\overline{)x^2 + 9x + 14} \\
\underline{(-)x^2 + (-)2x} \quad \text{Subtract} \\
7x + 14
\end{array}
$$

Step 2:

$$
\begin{array}{r}
x + 7 \qquad \text{Multiply } 7(x+2) \\
x+2\,\overline{)x^2 + 9x + 14} \\
\underline{(-)x^2 + (-)2x} \\
7x + 14 \\
\underline{(-)7x + (-)14} \\
0
\end{array}
$$

Check Example #1:

$$45$$
$$\underline{\times 14}$$
$$180$$
$$\underline{45}$$
$$630$$

Check Example #2

$$x + 7$$
$$\underline{x + 2}$$
$$x^2 + 7x$$
$$\underline{+ \quad 2x + 14}$$
$$x^2 + 9x + 14$$

THEY BOTH CHECK OUT !

Example #3:

Simplify: $\dfrac{6x^2 + x - 12}{3x - 4}$

Solution *

$$\begin{array}{r} 2x + 3 \\ 3x - 4 \overline{\smash{\big)}\, 6x^2 + x - 12} \\ \underline{(-)6x^2 - (+)8x} \\ 9x - 12 \\ \underline{(-)\, 9x - (+)12} \\ 0 \end{array}$$

<u>Check</u>

$$2x + 3$$
$$\underline{3x - 4}$$
$$6x^2 + 9x$$
$$\underline{\quad - 8x - 12}$$
$$6x^2 + x - 12$$

Example #4:

Simplify: $\dfrac{x^2 + 8x + 10}{x+5}$

Solution:

$$\begin{array}{r} -5 \\ x + 3 + \overline{x + 5} \\ x + 5 \overline{\smash{\big)}\, x^2 + 8x + 10} \\ \underline{(-)x^2 + (-)5x} \\ + 3x + 10 \\ \underline{(-)3x + -15} \\ -5 \end{array}$$

or $x + 3 - \dfrac{5}{x + 5}$

Notice there is a remainder in this example. See how it is written in the answer.

<u>**Check:**</u>

$$x + 3$$
$$\underline{x + 5}$$
$$x^2 + 3x$$
$$\underline{\quad\quad 5x + 15}$$
$$x^2 + 8x + 15$$
$$\underline{\quad\quad\quad -5}$$
$$x^2 + 8x + 10$$

Add the remainder

21

Let's Practice #10: (Divide the following)

1) $\dfrac{x^2 + 4x + 3}{x + 3}$

2) $\dfrac{x^2 + 7x + 12}{x + 4}$

3) $\dfrac{x^2 - 5x - 36}{x + 4}$

4) $\dfrac{y^2 - 3y - 54}{y + 6}$

5) $\dfrac{y^2 + 4y + 3}{y + 1}$

6) $\dfrac{y^2 + 3y - 4}{y + 4}$

7) $\dfrac{a^2 + 2a + 1}{a + 1}$

8) $\dfrac{b^2 + 3b - 4}{b - 1}$

9) $\dfrac{6a^2 + 4a + 3}{3a - 1}$

10) $\dfrac{3c^2 + 8c + 4}{3c + 2}$

11) $\dfrac{2x^2 - 5x - 2}{2x + 1}$

12) $\dfrac{m^2 - 25}{m - 5}$

If you had trouble with #12, here is a look-alike:

$\dfrac{x^2 - 16}{x - 4}$

$$\begin{array}{r} x + 4 \\ x - 4 \,\overline{\smash{)}\, x^2 + 0x - 16} \\ \underline{(-)x^2 -(+)4x } \\ +4x - 16 \\ \underline{(-)\, 4x - (+)16} \\ 0 \end{array}$$

⟵——— Spread out*

TRANSFORMING EQUATIONS INVOLVING DIVISION

Rule: When the variable is sitting beside the number, we divide both sides of the equation by that particular number.

Example #1: $4x = 8$

Example #2 $3x = 9$

$\dfrac{4x}{4} = \dfrac{8}{4}$

$\dfrac{3x}{3} = \dfrac{9}{3}$

Answer: $x = 2$

$x = 3$

Let's Practice #11

Simplify:

1) $6x = 12$

2) $7t = -7$

3) $5a = -15$

4) $3z = 12$

5) $2x = -14$

6) $12b = -288$

7) $7d = 322$

8) $-12b = -288$

9) $-16k = 176$

10) $15y = -195$

11) $-11t = 132$

12) $15t = 225$

TRANSFORMING EQUATIONS INVOLVING MULTIPLICATION

Rule: When the variable is sitting over the number, we multiply both sides of the equation by that particular number.

Example #1: $\dfrac{x}{2} = 9$

$$\dfrac{2}{1}\left(\dfrac{x}{2}\right) = \dfrac{9}{1}\left(\dfrac{2}{1}\right)$$

Answer: $x = 18$

Example #2 $\dfrac{a}{5} = 15$

$a = 75$

NOTE: $\dfrac{2}{1}$ still means 2. If there is a negative such as $\dfrac{-x}{2}$ or $-\dfrac{x}{2}$, we use $\dfrac{-2}{1}$

Rule: Where there is a fraction sitting in front of the variable, we multiply both sides of the equation by the reciprocal of that particular fraction.

Examples: $\dfrac{1}{3}x = 5$ and $\dfrac{2}{5}x = 2$

$$\left(\dfrac{3}{1}\right)\dfrac{1}{3}x = \dfrac{5}{1}\left(\dfrac{3}{1}\right)$$

$x = 15$

$$\left(\dfrac{5}{2}\right)\dfrac{2}{5}x = \dfrac{2}{1}\left(\dfrac{5}{2}\right)$$

$x = 5$

Let's Practice #12:

1) $\dfrac{1}{3}x = 2$

2) $\dfrac{1}{2}z = 6$

3) $\dfrac{-1}{5}t = -3$

4) $\dfrac{1}{4}w = \dfrac{3}{2}$

5) $\dfrac{-1}{21}x = -5$

6) $\dfrac{1}{9}x = 12$

7) $\dfrac{1}{6}a = 3$

8) $\dfrac{1}{4}t = 12$

9) $\dfrac{1}{7}t = 4$

10) $\dfrac{2}{3}x = 8$ 11) $\dfrac{3}{4}x = 12$ 12) $\dfrac{2}{5}t = 10$

SOLVING EQUATTONS USING SEVERAL TRANSFORMATIONS

Before we start, let's establish some standards by which we will work. The first thing we want to do is to define what we mean by the standard form of an equation. An equation is considered to be in standard form if it has the variables on the left and the constants on the right. For example:

<u>The Standard Form of an Equation</u>

Left		Right
x	=	some number

Example #1: $x + 3 = 7$

The first thing we want to do is to look for what is out of place. Since the constant 3 is out of place, and it is positive, we must move it by bringing in its opposite on both sides:

$$x + 3 = 7$$
$$x + 3 - 3 = 7 - 3$$

Answer: $x = 4$

Example #2: $y - 2 = 6$

$$y - 2 + 2 = 6 + 2$$

Answer: $y = 8$

Example #3: $2x + 3 = 11$

$$2x = 3 - 3 = 11 - 3$$
$$2x = 8$$
$$\dfrac{2x}{2} = \dfrac{8}{2}$$

Answer: $x = 4$ (Hint: Look back at solving equations by division.)

Example #4: $\dfrac{y}{5} + 2 = -10$

$$\dfrac{y}{5} + 2 - 2 = -10 - 2$$ Hint: $(-10 + -2) = -12$

$$\dfrac{y}{5} = -12$$ Hint: $\left(\dfrac{y}{5} \text{ means } \dfrac{1}{5}y\right)$

$$\left(\dfrac{5}{1}\right) \dfrac{y}{5} = \dfrac{-12}{1} \left(\dfrac{5}{1}\right)$$

Answer: $y = -60$

By replacing the variable with the answer we got, we can check the problem to see if it is correct. Always write the equation down again before beginning the check.

$$\frac{y}{5} + 2 = -10$$

$$\frac{-60}{5} + 2 = -10$$

$$-12 + 2 = -10$$

$$-10 = -10$$

Example #5: (Hint: -5 -5 means -5 + -5 = -10)

	Check
13 - 2(3t - 1) = -15	13 - 2(3t - 1) = -15
13 - 6t + 2 = -15	13 - 2(3 • 5 - 1) = -15
-6t + 15 = -15	13 - 2(15 -1) = -15
-6t + 15 - 15 = -15 -15	13 - 2(14) = -15
-6t = -30	13 - 28 = -15
$\frac{-6t}{-6} = \frac{-30}{-6}$	-15 = -15

Answer: t = 5

NOTE: When you see two or more negatives going across, add them together. Remember the distributive property. Remember the sign rules for addition, multiplication, and division of real numbers.

Example #6:

	Check
5(a + 2) = -25	5(a + 2) = -25
5a + 10 = -25	5 [(-7) + 2] = -25
5a + 10 - 10 = -25 -10	5(-5) = -25
5a = -35	-25 = -25
$\frac{5a}{5} = \frac{-35}{5}$	

Answer: a = -7

Let's Practice #13:

1) x + 2 = 5

2) x - 6 = 15

3) y + 4 = 7

4) 3x - 1 = 14

5) 4x - 13 = 11

6) 5x + 6 = 21

7) 12 + 5a = -78

8) 2y - 5 = 11

9) 3y - 2 = 7

10) c/5 + 20 = 23

11) 3(x + 1) + 2 = -7

12) 13 + 4(y + 5) = -3

SOLVING EQUATIONS WITH SIMILAR TERMS

Before we start, let's think back to the section on combining like or similar terms. Also, remember the sign rules for addition, subtraction, multiplication and division of real numbers.

Example #1: $3x + 5x - 2 = 46$

Combine like terms: $8x - 2 = 46$

Add +2 to both sides: $8x - 2 + 2 = 46 + 2$

Write the results: $8x = 48$

Divide both sides by 8: $\dfrac{8x}{8} = \dfrac{48}{8}$

Answer: $x = 6$

Example #2: $4(x + 1) - 5x = -3$

Clear the parentheses: $4x + 4 - 5x = -3$ (Distributive Axiom)

Combine like terms: $-x + 4 = -3$

Bring -4 on both sides: $-x + 4 - 4 = -3 - 4$

Write the results: $-x = -7$

Multiply both sides by -1: $(-1)(-x) = (-7)(-1)$ **NOTE:** $(-) \times (-) = +$

Answer: $x = 7$

NOTE: Whenever we end a problem with a negative variable equaling any number, the number can be negative or positive, but the variable itself must be positive. When this situation occurs, we must divide, or multiply, both sides of the equation by a -1 as shown in Example #2.

Let's Practice #14:

1) $2x + x = 18$ 2) $5y - y = 12$ 3) $-7z + 5z = 16$

4) $0 = m - 14 - 3m$ 5) $0 = n + 5 + 4n$ 6) $3(a + 2) - 4a = -9$

7) $2(a - 3) + (a + 3) = 9$ 8) $3(y + 2) - (y - 1) = 17$

9) $2(x - 1) - x = -2$ 10) $4(a + 2) - 2(1 - a) - 4a = 0$

11) $5(b - 2) - 2(b + 4) = 6$ 12) $5a + 2 [3(1-a) - 2(1 + a)] = 12$

If you had trouble with #4, #5, #8, and/or #12 here are some look-alikes.

Look-A-Like #5: $0 = x + 16 + 7x$

$0 = 8x + 16$

Let's recall the symmetric property: $a = b \implies b = a$

(Therefore) $8x + 16 = 0$

$8x + 16 - 16 = 0 - 16$

$8x = -16$

$\dfrac{8x}{8} = \dfrac{-16}{8}$

Answer: $x = -2$

Look-A-Like #8: $2(y + 3) - (y - 2) = 3$

Clear the parentheses: $2y + 6 - y + 2 = 3$ (Distributive Axiom)

$y + 8 = 3$

$y + 8 - 8 = 3 - 8$ Add $(+) + (-)$

Answer: $y = -5$

Check: $2(y + 3) - (y - 2) = 3$

$2(-5 + 3) - (-5 - 2) = 3$ [Add $(-) + (-)$]

$2(-2) - (-7) = 3$ [Multiply $-(-) = +$]

$-4 + 7 = 3$

$3 = 3$

Look-A-Like #12: $3a + 2[5(2-a) - 4(1 + a)] = -18$

Move the parentheses: $3a + 2[10 - 5a - 4 - 4a] = -18$

Combine inside brackets:	$3a + 2[-9a + 6] = -18$
Move the brackets:	$3a - 18a + 12 = -18$
Combine similar terms:	$-15a + 12 = -18$
Bring -12 on both sides:	$-15a + 12 - 12 = -18 - 12$ (Combine)
	$-15a = -30$
	$\dfrac{-15a}{-15} = \dfrac{-30}{-15}$
Answer:	$a = 2$

Solving Equations Having The Variable on Both Sides

Remember the standard form of an equation.

<u>Left</u>		<u>Right</u>
Variables	=	Constants (numbers)

NOTE: The variable appears on both sides of the equation.

Example #1:   Solve $3x = x + 14$

Because variables represent numbers, you may transform an equation by adding a variable expression to both sides or by subtracting a variable expression from both sides.

Let's see how it works using Example #1:

	<u>Check</u>
$3x = x + 14$	$3x = x + 14$
$3x - x = x - x + 14$	$3(7) = 7 + 14$
$2x = 14$	$21 = 21$
$\dfrac{2x}{2} = \dfrac{14}{2}$	
$x = 7$	

Example #2: $2(a - 5) = a - 1$ <u>Check</u>

$2a - 10 = a - 1$ $2(a - 5) = a - 1$

$2a - 10 + 10 = a - 1 + 10$ $2(9 - 5) = 9 - 1$

$2a = a + 9$ $2(4) = 8$

$2a - a = a - a + 9$ $8 = 8$

Answer: $a = 9$

Let's Practice #15:

1) $2y - 7 = 3y$ 2) $x + 3 = 1 - x$

3) $4(x - 1) = x + 8$ 4) $3(a + 1) = 2a$

5) $4a = 3a + 7$ 6) $13a = 40 + 8a$

7) $10y = -4 + 12y$ 8) $42 + 10y = 4y$

9) $-4y - 18 = -13y$ 10) $6(x + 8) = 4x$

11) $4(2x - 1) + 5 = 3x + 1$ 12) $3(y - 4) - y = 4(y - 2)$

MONOMIAL FACTORS OF POLYNOMIALS

When we talk about monomial factors, we want to place emphasis on the greatest monomial factor of a polynomial which is the greatest common factor of its terms. Notice the factoring in the following examples:

Example #1:   Factor: $\underline{8}a^3b - \underline{12}ab^3$

1.   First, let's look at the numerical coefficients "8" and "12". The greatest common factor of 8 and 12 is 4. Pull out the 4 and we have 4(   ).

2.   Next, we look to see if there is a variable part that is common. Yes, there is some form of a and b. Pick the if "a" with the lowest exponent, do the same for "b".

3.   This will be ab, now we have 4ab(   ).

4.   Now, to determine the other factor that goes inside of the parentheses, we must recall the rules we followed in the division of a polynomial by a monomial. Since we know the greatest monomial factor is 4ab, we will divide the original expression, $8a^3b - 12ab^3$ by 4ab:

$$\frac{8a^3b - 12ab^3}{4ab} = \frac{8a^3b}{4ab} - \frac{12ab^3}{4ab}$$

$$= \frac{8}{4} \cdot \frac{a^3}{a} \cdot \frac{b}{b} - \frac{12}{4} \cdot \frac{a}{a} \cdot \frac{b^3}{b}$$

$$= 2 \cdot a^2 \cdot 1 - 3 \cdot 1 \cdot b^2$$

$$= 2a^2 - 3b^2$$

5. Therefore, the answer to Example #1 is $4ab(2a^2 - 3b^2)$.

Example #2: Factor: $6x^3 - 9x^2 + 3x$

1. Look at the numerical coefficients (6, 9, 3). 3 is the greatest common factor.
   Take out "x" with the lowest exponent showing.

2. Look at the common variable in each term. "x" is the common variable.

3. Now, we pull out 3x and divide $6x^3 - 9x^2 + 3x$ by 3x.

$$\frac{6x^3 - 9x^2 + 3x}{3x} = \frac{6x^3}{3x} - \frac{9x^2}{3x} + \frac{3x}{3x}$$

Note: Remember anything written over itself is equal to 1.

$$= \frac{6}{3} \cdot \frac{x^3}{x} - \frac{9}{3} \cdot \frac{x^2}{x} + \frac{3x}{3x}$$

$$= 2 \cdot x^2 - 3 \cdot x + 1$$

$$= 2x^2 - 3x + 1$$

4. Therefore the answer to Example #2 is $3x(2x^2 - 3x + 1)$. After practicing, it won't be long before you will be able to do these types mentally.

Example #3: Factor: $4x^3 + 8x^2$

Here is the way we would do Example #3 mentally:

1. Look at the numerical coefficients "4" and "8". "4" is the greatest common factor.

2. Look for a common variable. "x^2" is the lowest exponent showing.

3. Pull out $4x^2(\ )$. Now, think of what you need.

4. You need $4x^2(x)$ to return to the value of $4x^3$.

5.   You need $4x^2(2)$ to return to the value of $8x^2$.

6.   Therefore, the answer to Example #3 is $4x^2(x + 2)$.

Let's Practice #16

1) $6a + 12a^2$ 2) $15b^2 - 9b$ 3) $5a^2 + 10a$

4) $x^2 - 4xy$ 5) $3h^2k + 3h$ 6) $2x + 6$

7) $2x^2 + 10xy$ 8) $12a^2b + 15ab^2$

9) $6y^4 + 9y^3 - 12y^2$ 10) $35t^2 + 25t^3 - 15t^4$

11) $a^3x^3 + a^2x^2 - 2ax$ 12) $4x^2 + 4x - 6$

FACTORING THE DIFFERENCE OF SQUARES

While remembering the sign rules for multiplying polynomials, using the vertical method, let's checkout what happens when we simplify the product $(a - b)(a + b)$:

$$
\begin{array}{r}
a + b \\
\underline{a - b} \\
a^2 + ab \\
\underline{ - ab - b^2} \\
a^2 + 0 - b^2
\end{array}
$$

         $\underline{a - b}$ ◄——————— Remember the sign rules for multiplication.

         $\underline{-ab - b^2}$ ◄——————— Remember the addition sign rules.

Answer: $a^2 - b^2$

Now, as you notice, $a^2 - b^2$ represents the difference of two squares. ("a" being squared and "b" being squared.) Since the product of $(a - b)(a + b) = a^2 - b^2$, then we can assume that $a^2 - b^2 = (a - b)(a + b)$. This is supported by our symmetric property which states: $a = b \Longrightarrow b = a$

$$(a - b)(a + b) = a^2 - b^2 \quad \Longrightarrow \quad a^2 - b^2 = (a - b)(a + b)$$

NOTE: $\Longrightarrow$ means "implies"

Let's keep this in mind as we look at some examples.

Example #1:   Factor: $a^2 - 4$

If this expresses a difference of two squares, what are they?

Well, let's see: $a^2 - 4$

                   $(a)^2 - (2)^2$ "a" represents one term that will be squared and 2 is the other.

Now, Knowing $(a)^2 - (b)^2 = (a - b)(a + b)$, we can conclude that $(a)^2 - (2)^2 = (a - 2)(a + 2)$,

∴   (Therefore) $a^2 - 4 = (a - 2)(a + 2)$ is the answer.

In short, we want to look at the square root of each term and determine if there exists a difference of two squares. If there does not exist a difference of two squares, then we can not factor as the difference of squares.

Square root - The second root of a particular quantity.

NOTE: (a) $\sqrt{4} = 2$ because $2 \times 2 = 4$

(b) $\sqrt{16} = 4$ because $4 \times 4 - 16$

(c) $\sqrt{x^2} = x$ because $x \cdot x = x^2$

(d) $\sqrt{x^4} = x^2$ because $x^2 \cdot x^2 = x^4$

(e) $\sqrt{9x^4} = 3x^2$ because $(3x^2)(3x^2) = 9x^4$

Example #2:   Factor: $4y^2 - 81$

Step 1: $4y^2 - 81$

Step 2: $\sqrt{4y^2} - \sqrt{81}$

Step 3: $2y - 9$

Step 4: $(2y)^2 - (9)^2 - (2y - 9)(2y + 9)$

Step 5:   ∴ $4y^2 - 81 = (2y - 9)(2y + 9)$ ◄——————— Answer

32

NOTE: To get the perfect square root of any variable, it must have an even exponent. To do so, you must get half of the exponent.

$\sqrt{x^6} = x^3$ because $x^3 \cdot x^3 = x^6$

Let's check our answer to Example #2:

$$2y - 9$$
$$\underline{2y + 9}$$
$$4y^2 - 18y$$
$$\underline{ + 18y - 81}$$
$$4y^2 + 0 - 81 = 4y^2 - 81$$

Let's Practice #17:

1) $a^2 - 1$ 2) $x^2 - 16$ 3) $y^2 - 9$

4) $b^4 - 49$ 5) $25x^2 - 9$ 6) $36a^2 - 16$

7) $16a^2 - b^2$ 8) $r^6 - t^6$ 9) $4x^2 - y^4$

10) $49m^4 - n^6$ 11) $a^{2n} - 36$ 12) $x^{4n} - 9y^{2n}$

If you had trouble with #11 and #12, here are some look-alikes:

Look-A-Like #11: $a^{2n} - 16$

$$\sqrt{a^{2n}} - \sqrt{16}$$

$$a^n - 4$$

NOTE: $x^n \cdot x^n = x^{n + n} = x^{2n}$

Rules for exponents:

$$\sqrt{x^{2n}} = x^n$$

$$(a^n)^2 - (4)^2 = (a^n - 4)(a^n + 4)$$

$$\therefore \quad a^{2n} - 16 = (a^n - 4)(a^n + 4)$$

$\underline{\text{Check}}$
$$a^n - 4$$
$$\underline{a^n + 4}$$
$$a^{2n} - 4a^n$$
$$\underline{\phantom{a^{2n}} + 4a^n - 16}$$
$$a^2n + 0 - 16 \qquad = \qquad a^{2n} - 16$$

33

Look-A-Like #12: $y^{4k} - 9x^{2k}$

$\sqrt{y^{4k}} - \sqrt{9x^{2k}}$

$y^{2k} - 3x^{k}$

> NOTE: Half of 4k is 2k
>
> and half of 2k is k.
>
> $x^{2k} \cdot x^{2k} = x^{4k}$
>
> $x^{k} \cdot x^{k} = x^{2k}$

$(y^{2k})^{2} - (3x^{k})^{2} = (y^{2k} - 3x^{k})(y^{2k} + 3x^{k})$

$\therefore (y^{4k} - 9x^{2k} = (y^{2k} - 3x^{k})(y^{2k} + 3x^{k})$

<u>Check</u>

$$
\begin{array}{l}
y^{2k} - 3x^{k} \\
\underline{y^{2k} + 3x^{k}} \\
y^{4k} - 3x^{k}y^{2k} \\
\underline{\quad\quad + 3x^{k}y^{2k} - 9x^{2k}} \\
y^{4k} \quad + 0 \quad\quad - 9x^{2k} = y^{4k} - 9x^{2k}
\end{array}
$$

FACTORING TRINOMIALS

At this time, we want to focus in on those types of trinomials which have "1" as the numerical coefficient of the first term.

Example #1: $x^{2} + 9x + 14$

When we factor a trinomial of this type, we factor the first term and the last term. When we factor the last term, we check its factor by adding them together. If the sum of the last term's factors is equal to that of the middle term numerical coefficient with the correct sign, the factors are correct for that particular trinomial.

Now, let's factor the given example.

(x) (x) → factors x^{2} → $x^{2} + 9x + 14$ ⟵ Write all of the
possible factors for
(x   )(x   )   +14.

NOTE: $x \cdot x = x^{2}$

<u>Possible Factors</u>	<u>Add Factors</u>
+14 = (2) (7)	2 + 7 = 9
= (-2) (-7)	-2 + -7 = -9
= (1) (14)	1 + 14 = 15
= (-1) (-14)	-1 + -14 = -15

Since the numerical coefficient of the middle term is +9, and the sum of the factors +2 and +7 is also +9, we conclude that +2 and +7 are the correct set of factors for the last term of this equation.

Therefore: $x^2 + 9x + 14$

Answer: $(x + 7)\ (x + 2)$

Now, let's check our answer:

$$(x + 7)\ (x + 2) \rightarrow$$

$$
\begin{array}{r}
x + 7 \\
\underline{x + 2} \\
x^2 + 7x \\
\underline{\quad 2x + 14} \\
x^2 + 9x + 14
\end{array}
$$

NOTE: This product gives us back the original equation. When this happens, we know that we have factored correctly.

Example #2: $x^2 - 5x - 14$

$(x\quad)(x\quad)$

Possible Factors	Add Factors
$-14 = (-7)(+2)$	$-7 + 2 = -5$
$= (+7)\ (-2)$	$7 + -2 = +5$
$= (+1)\ (-14)$	$1 + -14 = -13$
$= (-1)\ (+14)$	$-1 + 14 = +13$

The set of factors whose sum is -5 is (-7) and (+2). Therefore, $x^2 - 5x - 14 = (x - 7)(x + 2)$. Now let's check:

$$
\begin{array}{r}
x\ -\ 7 \\
\underline{x\ +\ 2} \\
x^2\ -\ 7x \\
\underline{\quad 2x\ -\ 14} \\
x^2\ -\ 5x\ -\ 14
\end{array}
$$

The answer equals the equation.

Let's Practice #18:

1) $x^2 + 7x + 6$ 2) $x^2 + 6x + 8$ 3) $y^2 - 3y + 2$

4) $a^2 + 2a - 8$ 5) $b^2 - b - 6$ 6) $c^2 - 5c - 24$

7) $y^2 + 9y + 20$ 8) $y^2 + 7y + 10$ 9) $y^2 + 6y - 27$

10) $d^2 - 14d - 32$ 11) $a^2 + 2ab + b^2$ 12) $a^2 - 14ab + 49b^2$

If #11 and #12 gave you trouble, here are some look-alikes:

Look-A-Like #11:

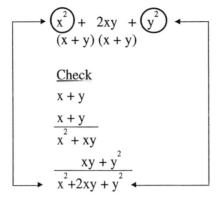

$$x^2 + 2xy + y^2$$
$$(x + y)(x + y)$$

Check

$x + y$

$\dfrac{x + y}{x^2 + xy}$

$\dfrac{xy + y^2}{x^2 + 2xy + y^2}$

Look-A-Like #12

$$x^2 - 12xy + 36y^2$$
$$(x - 6y)(x - 6y)$$

Check

$x - 6y$

$\dfrac{x - 6y}{x^2 - 6xy}$

$\dfrac{- 6xy + 36y^2}{x^2 - 12xy + 36y^2}$

FACTORING TRINOMIALS

At this time, we want to focus on those trinomials which have numerical coefficients greater than 1 in the first term.

Example #1: $2x^2 + 5x + 3$

$(2x + 3)(x + 1)$

On this type of trinomial, we must use the trial and error method. Because of the 2 sitting in front of the variable in the first term, we can not factor this type of trinomial in the same manner as we factored a trinomial with 1 as its numerical coefficient.

NOTE: Keep in mind that there are some look-alikes following each exercise. Please make note and take advantage of them.

Remember
Don't be too impatient, take time to read through every step of your examples and look-alikes.

_____

Since we are going to use a trial and error method, let's use a quick check to make things easier. Only the correct factors will give the correct middle term.

NOTE: Multiply the two inside terms and the two outside terms, then add the results for a quick check for the middle term.

Example #2: $3x^2 + 2x - 8$ Possible Factors

$$(3x - 2)(x + 4)$$ $-8 = (-2)(4)$

- 2x		$= (2)(-4)$
+ 12x	Add	$= (1)(-8)$
+ 10x	Reject	$= (-10)(8)$

Let's try again: $3x^2 + 2x - 8$ **NOTE:** Try switching the signs around sometimes.

$$(3x + 2)(x - 4)$$

+ 2x	
- 12x	Add
- 10x	Reject

Let's try again: $3x^2 + 2x - 8$ This time we will switch our factors.

$$(3x - 4)(x + 2)$$

- 4x	
+ 6x	Add
+ 2x	**Correct**

THE MIDDLE TERM CHECKS OUT. Therefore the correct factors are +2 and -4. Let's check Example #2 with the vertical method and the quick check to compare:

$$
\begin{array}{ll}
3x - 4 & \\
\underline{x + 2} & \\
3x^2 - 4x & \\
\underline{\quad 6x - 8} & \\
3x^2 + 2x - 8 &
\end{array}
$$

$(3x - 4) (x + 2)$

-4x

+6x

+2x ⟵ (middle term)

If we use the quick check and we can obtain the middle term with the correct sign, we know that the factors we selected are correct.

Let's Practice #19   (Factors each)

1) $2a^2 - 5a - 3$ 2) $3a^2 + 13a - 10$

3) $2a^2 + 7a + 3$ 4) $5x^2 + 11x + 2$

5) $4x^2 - 8x - 5$ 6) $18x^2 - 19x - 12$

7) $36c^2 - 5c - 24$ 8) $12c^2 - 4c - 21$

9) $24c^2 - 31c - 15$ 10) $y^2 - xy - 56x^2$

11) $6 - 23y - 4y^2$ 12) $9y^2 + 3y - 2$

SPECIAL CASES OF FACTORING

We have practiced factoring trinomials and the difference of squares. Now let's look at some special factoring of these types:

$$\text{Factor: } (x + 1)^2 + 9(x + 1) + 14$$

Before we start, let's think back and review what we did when we were factoring trinomials. Our first example was $x^2 + 9x + 14$ and we factored it as follows:

$$x^2 + 9x + 14$$
$$(x + 7)(x + 2)$$

Now, knowing that we can factor a situation such as this, what is the problem with factoring $(x + 1)^2 + 9(x + 1) + 14$? Well, in most cases a person will say that the parentheses are causing the problem. If that's so with you, let's suppose that the parentheses are something else by substitution:

$$(x + 1)^2 + 9(x + 1) + 14$$

1.   Let $(x + 1) = a$
2.   Then we have $(a)^2 + 9(a) + 14$
3.   Now factor: $a^2 + 9a + 14$
4. $(a + 7)(a + 2)$
5.   Since $(x + 1) = a$, substitute $(x + 1)$ for a.
6.   Use brackets: $[(x + 1) + 7]\ [(x + 1) + 2]$
7.   Simplify: $[x + 1 + 7]\ [x + 1 + 2]$
8.   Back to parentheses: $(x + 8)(x + 3)$ is the answer.

Let's check the answer to this special case.

Answer: $(x + 1)^2 + 9(x + 1) + 14 = (x + 8)(x + 3)$

If the answer is correct, both sides should be equal, once simplified:

$(x + 1)^2 + 9(x + 1) + 14 = (x + 8)(x + 3)$

$$(x + 1)^2 = (x + 1)(x + 1) \quad \Big| \quad 9(x+1) = 9x + 9 \Big| + 14 \qquad\qquad (x + 8)(x + 3)$$

$$
\begin{array}{ll}
x + 1 & \qquad\qquad x + 8 \\
\underline{x + 1} & \qquad\qquad \underline{x + 3} \\
x^2 + x & \qquad\qquad x^2 + 8x \\
\underline{\quad\quad x + 1} & \qquad\qquad \underline{\quad\quad 3x + 24} \\
x^2 + 2x + 1 & \qquad\qquad x^2 + 11x + 24
\end{array}
$$

$$x^2 + \underline{2x} + \underline{1} + \quad \Big| \quad \underline{9x} + \underline{9} + \underline{14} \quad \Big| \quad =$$

(Checks out) $x^2 + 11x + 24 = x^2 + 11x + 24$

Let's Practice #20: Factor:

1) $(a + b)^2 - (a + b) - 6$

2) $(x + 1)^2 - 2(x + 1) - 48$

3) $(c + d)^2 - 5(c + d) - 6$

4) $(x - 2)^2 + 4(x - 2) - 60$

Now let's look at the trinomial that we used to experiment with the trial and error method. Take a few minutes to review this section of the book.

Factor: $2x^2 + 5x - 12$
$(2x - 3)(x + 4)$

Let's compare our next example.

Example #1: $2(x + 1)^2 - (x + 1) - 3$

1.   Let $(x + 1) = a$
2.   Then we have $2(a)^2 - (a) - 3$
3.   Now factor: $2a^2 - a - 3$
4. $(2a - 3)(a + 1)$
5.   Since $(x + 1) = a$, substitute $(x + 1)$ for a.
6.   Use brackets: $[2(x + 1) - 3][(x + 1) + 1]$
7.   Simplify: $[2x + 2 - 3][x + 1 + 1]$
8.   Replace parentheses: $(2x - 1)(x + 2)$

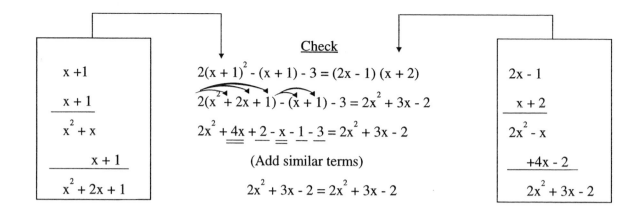

Check

$2(x + 1)^2 - (x + 1) - 3 = (2x - 1)(x + 2)$

$2(x^2 + 2x + 1) - (x + 1) - 3 = 2x^2 + 3x - 2$

$2x^2 + 4x + 2 - x - 1 - 3 = 2x^2 + 3x - 2$

(Add similar terms)

$2x^2 + 3x - 2 = 2x^2 + 3x - 2$

Left box:

x + 1
x + 1
_____
$x^2 + x$
        x + 1
_____
$x^2 + 2x + 1$

Right box:

2x - 1
x + 2
_____
$2x^2 - x$
        +4x - 2
_____
$2x^2 + 3x - 2$

Let's Practice #21:   Factor:

1) $4(x + y)^2 - 5(x + y) - 6$ 2) $3(x + y)^2 - 17(x + y) - 6$

3) $2(a + 3)^2 + 5(a + 3) - 3$ 4) $5(a + 1)^2 - 2(a + 1) - 7$

SPECIAL CASES OF FACTORING THE DIFFERENCE OF SQUARES

Let's think back to the practice we did on factoring $a^2 - b^2$.

Given: $a^2 - b^2 - = (a + b)(a - b)$

Example #1: $x^2 - 4$
                          $\sqrt{x^2} - \sqrt{4}$
                          x - 2

Answer: (x - 2)(x + 2)

Remember x^2 represents x being squared and 4 represents 2 being squared.

Example #2: $x^2 - 36$
                          $\sqrt{x^2} - \sqrt{36}$
                          x - 6

Answer: (x - 6)(x + 6)

Let's look at a special case:

Example 3: $(a + 3)^2 - a^2$

1.   Let $(a + 3) = b$

2.   Then we have $(b)^2 - a^2$

3.   Factor: $b^2 - a^2$

4. $(b - a)(b + a)$

5.   Since $(a + 3) = b$, substitute $(a + 3)$ for b.

6.   Use brackets: $[(a + 3) - a][(a + 3) + a]$

7.   Simplify: $[a + 3 - a][a + 3 + a]$

8.   Re-insert parentheses: $(3)(2a + 3)$

9.   Leave the 3 outside: $3(2a + 3)$ is the answer.

Example #4: $(x - 1)^2 - (x + 1)^2$

1.   Let $(x - 1) = a$ and let $(x + 1) = b$, since we have two different parentheses.

2.   Then we have: $(a)^2 - (b)^2$

3.   Factor: $a^2 - b^2$

4. $(a - b)(a + b)$

5.   Since $(x - 1) = a$, substitute $(x - 1)$ for a. Also, since $(x + a) = b$, substitute $x + 1$ for b.

6.   Use brackets: $[(x - 1) - \overparen{(x + 1)}][(x - 1) + \overparen{(x + 1)}]$

NOTE: (-) times everything inside the parentheses.

7.   Simplify: (Add like terms) $[\underline{x} - \underline{1} - \underline{x} - \underline{1}] \, [\underline{x} - \underline{1} + \underline{x} + \underline{1}]$

8.   Back to the parentheses: $(-2)(2x)$

9.   Write the product $-4x$ as the answer.

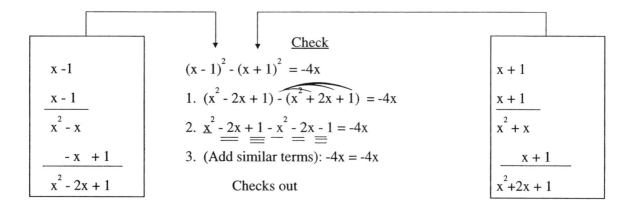

Let's Practice #22:   Factor:

1) $(a + 1)^2 - a^2$ 2) $x^2 - (x - 1)^2$

3) $(x + 2)^2 - (x + 3)^2$ 4) $9(x + 1)^2 - 4(x - 2)^2$

If you had trouble with #3 and 4, here are some look-alikes:

Look-A-Like #3: $(x + 4)^2 - (x + 5)^2$

1.   Let $(x + 4) = a$, and $(x + 5) = b$

2.   Then we have $(a)^2 - (b)^2$

3.   Factor: $a^2 - b^2$

4. $(a - b)(a + b)$

5.   Since $(x + 4) = a$ and $(x + 5) = b$, substitute $(x + 4)$ for a, and $(x + 5)$ for b.

6.   Use brackets: $[(x + 4) - (x + 5)][(x + 4) + (x + 5)]$

7.   Simplify: (Add) $[x + \underline{4} - x - \underline{5}][x + \underline{4} + x + \underline{5}]$

8.   Back to parentheses: $(-1)(2x + 9)$

9.   Your choice for the answer: $-1(2x + 9)$ or $-(2x + 9)$

If we move the parentheses by multiplying the negative sign all the way through, thus using the distributive property, the answer would appear as follows: $-1(2x + 9) = -2x - 9$

Look-A-Like #4: $16(x + 2)^2 - 4(x - 3)^2$

1.   Let $(x + 2) = m$ and let $(x - 3) = n$

2.   Then we have $16(m)^2 - 4(n)^2$

3.   Factor: $16m^2 - 4n^2$

4.   Square Root? $\sqrt{16m^2} - \sqrt{4n^2}$

5. $4m - 2n$

6. $(4m - 2n)(4m + 2n)$

7.   Since $(x + 2) = m$ and $(x - 3) = n$, substitute $(x + 2)$ for m and $(x - 3)$ for n

8.   Use brackets: $[4(x + 2) - 2(x - 3)] \, [4(x + 2) + 2(x - 3)]$

9.   Simplify: $[\underline{4x} + \underline{8} - 2x + \underline{6}] \, [\underline{4x} + \underline{8} + 2x - \underline{6}]$

10.   Back to parentheses for the answer: $(2x + 14)(6x + 2)$

Let's <u>**Check**</u>

$$16(x + 2)^2 - 4(x - 3)^2 = (2x + 14)(6x + 2)$$

Multiply on both sides $16(x^2 + 4x + 4) - 4(x^2 - 6x + 9) = (2x + 14)(6x + 2)$

IT CHECKS OUT $16x^2 + \underline{64x} + \underline{64} - 4x^2 + \underline{24x} - \underline{36} = 12x^2 + 88x + 28$

$$12x^2 + 88x + 28 = 12x^2 + 88x + 28$$

FACTORING BY GROUPING

In order to factor by grouping, we must look for common factors. Those terms with common factors should be grouped together.

Example #1: $12ac - 4b - 6a + 8bc$

1.   Re-group: $(12ac - 6a) + (8bc - 4b)$

   (Notice the greatest monomial factors are 6a and 4b)

2. $= 6a(2c - 1) + 4b(2c - 1)$

3.   **Answer:** $= (6a + 4b)(2c - 1)$

   $2(3a + 2b)(2c - 1)$

You may want to look at more than one way to group terms:

$12ac - 4b - 6a + 8bc$

$= (12ac + 8bc) + (- 6a - 4b)$

$= 4c\underset{\text{finger}}{(3a + 2b)} + (-2)\underset{\text{finger}}{(3a + 2b)}$

Answer: $= (4c - 2)(3a + 2b)$

   $2(2c - 1)(3a + 2b)$ or $(2c - 1)(6a + 4b)$

Put your pointing finger from each hand over the parentheses that have the common binomials. (They are indicated.) You will see $4c + (-2)$ or $4c - 2$. Write this in parentheses on the left and write the common binomial that is under your finger in a parentheses on the right as shown above.

43

Let's Practice #23:   Remember to pull out the greatest common factor on the left and on the right sides.

1) $2y + xy + 2z + xz$ 2) $a^2 + ab + ac + bc$

3) $2y^3 + y^2 + 8y + 4$ 4) $3ac - 6bc - 4a + 8b$

5) $2x^3 - x^2 - 8x + 4$ 6) $2a^3 + 6a^2 + 5a + 15$

If you had trouble with #4, look back at the example given with the negative signs involved. Always look for a way to have a left and a right side set-up: $(3ac - 6bc) + (-4 + 8b)$

If you had trouble with #5, here is a look-alike:

$$x^3 - 2x^2y - 4x + 8y$$

$$(x^3 - 2x^2y) + (-4x + 8y)$$

Pull out the greatest monomial factors:

$x^2(x - 2y) + (-4)(x - 2y)$ or $x^2(x - 2y) - 4(x - 2y)$

Use the two index finger technique as we did previously:

$(x^2 - 4) (x - 2y)$ NOTE: $a^2 - b^2 = (a + b)(a - b)$ Since $x^2 - 4$ (The difference of squares) the answer is:

$$(x + 2) (x - 2)(x - 2y)$$

Check (Vertical Method)

Step 1	Step 2
$x - 2$	$x^2 - 4$
$\underline{x + 2}$	$\underline{x - 2y}$
$x^2 - 2x$	$x^3 - 4x$
$\underline{\quad + 2x\ -4}$	$\underline{\qquad\qquad -2x^2y + 8y}$
$x^2\qquad\quad - 4$	$x^3 - 4x - 2x^2y + 8y$

Let's write the answer in a more standard form:

$x^3 - 2x^2y - 4x + 8y$

Now, let's look at some more expressions to factor using the index finger technique.

Example #1:   x(z + 2) + y(z + 2)

(Use your index fingers)   (x + y)(z + 2) is the answer

Let's Practice #24:

1)   x(a + b) + 2(a + b)   2)   3x(x + 1) + 2(x + 1)

3)   x(2y - 1) + y(2y - 1)   4)   a(b + 3) - (b +. 3)

5)   (x - 6) + y(x - 6)   6)   a(2 - x) + b(x - 2)

If you had trouble with #4 and #6, here are some look-alikes:

Look-A-Like #4:   x(y + 5) - (y + 5)

NOTE: If there is nothing written in front of a parenthesis, it is understood to be a one (1) in front of the parenthesis.

Now, let's rewrite the problem: x(y + 5) - 1(y + 5). Use the index finger technique. The answer is (x - 1)(y + 5)

Look-A-Like #6: 2a(3 - y) + 6(y - 3)

Notice the difference inside the parentheses. Let's factor out a negative one (-1) on the left side. We now have:

$$2a(-1)\,(-3 + y)\ \ \text{or}\ \ -2a(-3 + y)\ \ \text{or}\ -2a(y - 3)$$

(The latter being more standard to write.) Let's continue.

Now, -2a(y - 3) + 6(y - 3) Use the index finger technique. We have (-2a + 6) (y - 3) or (6 - 2a) (y - 3)

USING MORE THAN ONE METHOD OF FACTORING

A polynomial is said to be factored completely when it is expressed as the product of a prime polynomial and monomial. In other words, when the polynomial has been factored until it can not be factored any further.

Let's look at some guidelines for such factoring. Remember every way that you have learned to factor.

Guidelines for Factoring

1)   Always look for a common factor first. If it is there, pull it out.

2)   Look for the differences of two squares. If they exist, you've got to factor them.

3) If a trinomial square is present, factor it.

4) If a trinomial is not a square, look for a pair of binomial factors.

5) If a polynomial has four or more terms, look for a way to group the terms in pairs, using a left and right side.

6) Make sure that each factor is prime. Check by multiplying the factors.

NOTE: GMF means "the greatest monomial factor"

Example #1: Factor: $2xy^2 - 18xz^2$

1. Pull out the GMF: $2xy^2 - 18xz^2 = 2x(y^2 - 9z^2)$

2. Next look for the difference of two squares.

3. **Answer:** $= 2x(y - 3z)(y + 3z)$

Example #2: Factor: $-3x^2 + 18x - 27$

1. $-3x^2 + 18x - 27 = -3(x^2 - 6x + 9)$ **(GMF)**

   $= -3(x - 3)(x - 3)$ **(Binomial square)**

   $= -3(x - 3)^2$ **Answer**

Example #3: Factor $16ax^2 - 12ax - 18a$

1) $16ax^2 - 12ax - 18a = 2a(8x^2 - 6x - 9)$ **(GMF)**

   $= 2a(8x^2 - 6x - 9)$ **(Factor the trinomial)**

   $= 2a(2x - 3)(4x + 3)$ **Answer**

NOTE: Remember the trial and error method when the numerical coefficient of the first term is greater than 1. Go back and review this section, if necessary.

Let's Practice #25   Factor:

Remember, we can factor the difference of squares, but not the sum of squares.

1) $8x^2 - 8x + 2$ 2) $x^4 - 5x^2 + 4$

3) $2a^2 - 8b^2 + 16b - 8$ 4) $y^2 - 9z^2 - 2y + 6z$

5) $y^3 - y^2 - 4y + 4$ 6) $2a^4 - 6a^2 - 8$

If you had trouble working #2, #3, #4 and #6, here are some look-alikes:

Look-A-Like #2: $y^4 - 10y^2 + 9$ **(Trinomial)**

$\qquad\qquad\qquad$ $(y^2 - 9)(y^2 - 1)$ **(The difference of squares) (Answer)**

$\qquad\qquad$ $(y - 3)(y + 3)(y + 1)(y - 1)$ **(Answer)**

Let's <u>Check</u>

$y - 3$	$y + 1$	$y^2 - 9$
$\underline{y + 3}$	$\underline{y - 1}$	$\underline{y^2 - 1}$
$y^2 - 3y$	$y^2 + y$	$y^4 - 9y^2$
$\underline{+ 3y - 9}$	$\underline{- y - 1}$	$\underline{- 1y^2 + 9}$
$y^2 - 9$	$y^2 - 1$	$\boxed{y^4 - 10y^2 + 9}$

The answer is correct

Look-A-Like #3: $3z^2 - 3y^2 + 24y - 48$

$\qquad\qquad\qquad$ $3(x^2 - y^2 + 8y - 16)$

$\qquad\qquad\qquad$ $3[x^2 - (y^2 - 8y + 16)]$ (Factoring out the -1)

$\qquad\qquad\qquad$ $3(x^2 - (y - 4)(y - 4)1$ Thus forming a trinomial square

$\qquad\qquad\qquad$ $3[x^2 - (y - 4)^2]$ (Binomial square)

$\qquad\qquad$ Let $(y - 4) - a$ (Remember special cases of factoring)

Then we have: $3(x^2 - a^2]$ (Differences of squares)

$\qquad\qquad$ $3[(x - a)(x + a)]$

Next, substitute for a, since a - (y - 4).

$$3[x - (y - 4)\ x + (y - 4)$$

$$3(x - y + 4)(x + y - 4)$$ **Answer**

Let's <u>Check</u>

$$
\begin{array}{l}
x - y + 4 \\
\underline{x + y - 4} \\
x^2 - xy + 4x \\
xy - y^2 + 4y \\
\underline{ - 4x + 4y - 16} \\
\end{array}
$$

Multiply by 3 $3 (x^2 \qquad -y^2 + 8y - 16)$

The answer is correct: $3x^2 - 3y^2 + 24y - 48$

Look-A-Like #4: $a^2 - 4b^2 - 3a + 6b$

Look for left and right side grouping:

$\qquad$ $(a - 2b)(a + 2b) - 3(a - 2b)$ $\qquad$ Difference of squares

$\qquad$ $(a + 2b) - 3$ $\qquad$ (Use index finger technique)

$\qquad$ $(a + 2b) - 3)\,(a - 2b)$

<u>Check</u>

$$a + 2b - 3$$
$$\underline{a - 2b\qquad\quad}$$
$$a^2 + 2ab - 3a$$
$$\underline{\quad -2ab\qquad\quad -4b^2 + 6b}$$

The answer is correct: $\qquad$ $a^2 \qquad\quad -3a \quad -4b^2 + 6b$

Look-A-Like #6:

$$3x^4 - 24x^2 - 27$$
$$3(x^4 - 8x^2 - 9) \qquad \text{(GMF)}$$
$$3(x^2 + 1)(x^2 - 9) \qquad \text{(Difference of squares)}$$
$$3(x^2 + 1)(x - 3)(x + 3) \qquad \textbf{Answer}$$

The <u>Check</u> is left for you to do.

SOLVING EQUATIONS IN FACTORED FORM

NOTE: $\quad$ For all real numbers a and b: $ab = 0$ if and only if, $a = 0$ or $b = 0$.

Let's look at the Zero-Product property:

Example #1: $\qquad$ Solve $(x + 2)(x + 3) = 0$

One of the parentheses on the left side must be equal to 0.

$\qquad$ $x + 2 = 0$ $\qquad\qquad$ OR $\qquad\qquad$ $x + 3 + 0$
$\qquad$ $x + 2 - 2 = 0 - 2$ $\qquad\qquad\qquad\qquad$ $x + 3 - 3 = 0 - 3$
$\qquad$ $x = -2$ $\qquad\qquad\qquad\qquad\qquad\quad$ $x = -3$

The solution set $\left\{\ -2,\ -3\right\}$

This simply means that if x is replaced by -2, the parentheses (x + 2), would result as 0. If x is replaced by -3, the parentheses (x + 3), would result as 0. Zero times either parentheses will cause the entire expression to equal to 0.

Example #2: Solve $x(x - 5) = 0$

$$x = 0 \qquad \text{or} \qquad x - 5 = 0$$
$$x - 5 + 5 = 0 + 5$$
$$x = 5$$

Solution set $\left\{ \ 0, 5 \ \right\}$

Let's Practice #26: Solve:

1) $(x - 6)(x + 6) = 0$ 2) $(y - 7)(y - 8) = 0$

3) $3a(a - 7) = 0$ 4) $b(b - 2) = 0$

Sample: $(3x - 2)(2x + 1) = 0$

$$3x - 2 - 0 \qquad\qquad\qquad \text{or} \qquad\qquad\qquad 2x + 1 = 0$$

$$3x - 2 + 2 = 0 + 2 \qquad\qquad\qquad\qquad\qquad 2x + 1 - 1 = 0 - 1$$

$$3x - 2 \qquad\qquad\qquad\qquad\qquad\qquad\qquad 2x = -1$$

$$\frac{3x}{3} = \frac{2}{3} \qquad\qquad\qquad\qquad\qquad\qquad \frac{2x}{2} = \frac{-1}{2}$$

$$x = \frac{2}{3} \qquad\qquad\qquad\qquad\qquad\qquad\qquad x = -\frac{1}{2}$$

Solution set $\left\{ \dfrac{2}{3} \ - \ \dfrac{1}{2} \right\}$

5) $(3a + 1)(2a - 3) = 0$ 6) $(3a - 4)5a - 0$

7) $(3x + 2)(x - 3) = 0$ 8) $2a(3a - 5)(a + 2) = 0$

Solving Equations by Factoring

Since we are working with polynomials, let's start by defining the term "polynomial equation". A polynomial equation is an equation which has a polynomial on one side and 0 on the other side. Polynomial equations are named after the term of the highest degree.

$ay + b = 0$ Linear equation **(exponent of 1)**

$ay^2 + by + c = 0$ Quadratic equation **(exponent of 2)**

$ay^3 + by^2 + cy + d = 0$ Cubic equation **(exponent of 3)**

A lot of times, we solve polynomial equations by factoring and then apply the zero-product property. Earlier in this book we mentioned the standard form of an equation. Now, let's focus on the **standard form of a polynomial equation.** This is when one side is 0 and the other side is a simplified polynomial that is arranged in descending powers of the variable. Sometimes when we are factoring a polynomial equation it loses its standard form. When this occurs, we must re-write the equation into the standard form.

Out of standard form: $x^2 - 9x = 14$

Rewritten in standard form: $x^2 - 9x - 14 = 0$

If you cannot move the terms mentally, here is the way to show your work:

$x^2 - 9x = 14$ **(14 is out of place)**

$x^2 - 9x - 14 = 14 - 14$ **(Bring in its opposite on both sides)**

$x^2 - 9x - 14 = 0$ **(Answer)**

Example #1:   Solve the quadratic equation: $x^2 - 36 = 0$

(Remember the difference of squares)

1. $x^2 - 36 = 0$

2. $(x - 6)(x + 6) = 0$ **(Factor)**

3.   Now, from a previous lesson we know that either:

         $x - 6 = 0$ **OR** $x + 6 - 0$

4          $x - 6 + 6 = 0 + 6$ $x + 6 - 6 = 0 - 6$ **(solve)**

         $x = 6$ $x = -6$

5   Solution set. $\left\{ 6, -6 \right\}$

<u>Check</u>	<u>Check</u>
$x^2 - 36 = 0$	$x^2 - 36 = 0$
$(6)^2 - 36 = 0$	$(-6)^2 - 36 = 0$
$36 - 36 = 0$	$36 - 36 = 0$
$0 = 0$	$0 = 0$

(Remember GMF is "greatest monomial factor".)

Example #2:   Solve the quadratic equation: $3a^2 - 6a = 9$

$$3a^2 - 6a = 9$$

1. $3a^2 - 6a - 9 = 0$ (Rewritten in standard form)

2. $3(a^2 - 2a - 3) = 0$ (GMF)

3. $3(a - 3)(a + 1) = 0$ (Factor)

4. $3 \neq 0, a - 3 = 0$ OR $a + 1 = 0$ **(Reject $3 \neq 0$)**

5. $a - 3 + 3 = 0 + 3$ $a + 1 - 1 = 0 - 1$ **(Solve)**

         $a = 3$ $a = -1$

6.   Solution set $\left\{\ 3,\ -1\ \right\}$

<u>Check</u>	<u>Check</u>
$3a^2 - 6a = 9$	$3a^2 - 6a - 9$
$(3)^2 - 6(3) = 9$	$3(-1)^2 - 6(-1) = 9$
$3(9) - 18 = 9$	$3(1) + 6 = 9$
$27 - 18 = 9$	$3 + 6 = 9$
$9 = 9$	$9 = 9$

Example #3: Solve the cubic equation: $y^3 - 12y^2 + 32 = 0$

1. $y^3 - 12y^2 + 32y = 0$

2. $y(y^2 - 12y + 32) = 0$ **(GMF)**

3. $y(y - 8)(y - 4) = 0$ **(Factor)**

4. $y = 0$ **OR** $y - 8 = 0$ **OR** $y - 4 = 0$

Solution set: 0,   8,   4

Check	Check	Check
$y^3 - 12y^2 + 32y = 0$	$y^3 - 12y^2 + 32y = 0$	$y^3 - 12y^2 + 32y = 0$
$(0)^3 + 12(0)^2 + 32(0)^2 = 0$	$(8)^3 - 12(8)^2 + 32(8) = 0$	$(4)^3 - 12(4)^2 + 32(4) = 0$
$0 + 0 + 0 = 0$	$512 - 768 + 256 = 0$	$64 - 192 + 128 = 0$
$0 = 0$	$0 = 0$	$0 = 0$

Let's Practice #27:   Solve:

(Refer back to examples as needed.)

1) $x^2 - 16x + 63 = 0$ (2) $x^2 + 2x - 63 = 0$

3) $x^2 - 5x = 24$ 4) $4y^2 - 9 = 0$

5) $y^2 - 4y + 4 = 0$ 6) $3y^2 + 2y - 1 = 0$

7) $b^3 - b = 0$ 8) $b^3 - 4b = 0$

9) $2b^2 - 5b - 3 = 0$ 10) $a^3 + 49a = 14a^2$

11) $a^4 - 13a^2 + 36 = 0$ 12) $a^3 + a^2 = 4a + 4$

If you had trouble with #4, #7, and #12, here are some look-alikes:

Look-A-Like #4: $9y^2 - 16 = 0$

1. $(3y - 4)(3y + 4) - 0$ (Factor)

2. $3y - 4 = 0$ **OR** $3y + 4 = 0$ (Zero product property)

3. $3y - 4 + 4 = 0 + 4$ $3y + 4 - 4 = 0 - 4$

               $3y = 4$ $3y = -4$

               $\dfrac{3y}{3} = \dfrac{4}{3}$ $\dfrac{3y}{3} = \dfrac{-4}{3}$

               $y = \dfrac{4}{3}$ $y = \dfrac{-4}{3}$

Solution set $\left\{ \dfrac{4}{3}, \dfrac{-4}{3} \right\}$

Let's check Look-alike #4:

   $9y^2 - 16 = 0$ $9y^2 - 16 = 0$

   $9 \left(\dfrac{4}{3} \right)^2 - 16 = 0$ $9 \left(\dfrac{-4}{3} \right)^2 - 16 + 0$

   $\overset{1}{\cancel{9}} \left(\dfrac{16}{\underset{1}{\cancel{9}}} \right) - 16 = 0$ $\overset{1}{\cancel{9}} \left(\dfrac{16}{\underset{1}{\cancel{9}}} \right) - 16 = 0$

   $16 - 16 = 0$ $16 - 16 = 0$

   $0 = 0$ $0 = 0$

Look-A-Like #7: $x^3 - x - 0$

1. $x^3 - x = 0$

2. $x(x^2 - 1) = 0$ (Difference of squares)

3. $x(x - 1)(x + 1) = 0$ (Factor)

4. $x = 0$ **OR** $x - 1 = 0$ **OR** $x + 1 = 0$

        $x = 0$ $x - 1 + 1 = 0 + 1$ $x + 1 - 1 + 0 - 1$

5. $x = 0$ $x = 1$ $x = -1$

6.   Solution set   0, 1, -1

<u>Check</u>	<u>Check</u>	<u>Check</u>
$x^3 - x = 0$	$x^3 - x = 0$	$x^3 - x = 0$
$(0)^3 - 0 = 0$	$(1)^3 - 1 = 0$	$(-1)^3 - (-1) = 0$
$0 - 0 = 0$	$1 - 1 = 0$	$-1 + 1 = 0$
$0 = 0$	$0 = 0$	$0 = 0$

Remember: $(-1)^3 = -1 \cdot -1 \cdot -1 = -1$

     Also: $-(-1) = +1$

Look-A-Like #12: $b^3 + b^2 = 9b + 9$

1. $b^3 + b^2 = 9b + 9$

2. $b^3 + b^2 = 9b + 9$ (Move 9b)

            $b^3 + b^2 - 9b = 9$

3. $b^3 + b^2 - 9b = 9$ (Move 9)

            $b^3 + b^2 - 9b - 9 = 0$ (Standard form)

4. $b^2(b + 1) - 9(b + 1) = 0$ (GMF left and right sides)

5. $(b^2 - 9)(b + 1) = 0$ (Difference of squares)

6. $(b - 3)(b + 3)(b + 1) = 0$ (Factor)

7. $b - 3 = 0$ **OR** $b + 3 = 0$ **OR** $b + 1 + 0$

8. $b - 3 + 3 = 0 + 3$ $b + 3 - 3 = 0 - 3$ $b + 1 - 1 = 0 - 1$

            $b = 3$ $b = -3$ $b = -1$

9.   Solution set 3, -3, -1

Check	Check	Check
$b^3 + b^2 = 9b + 9$	$b^3 + b^2 = 9b + 9$	$b^3 + b^2 = 9b + 9$
$(3)^3 + (3)^2 = 9(3) + 9$	$(-3)^3 + (-3)^2 = 9(-3) + 9$	$(-1)^3 + (-1)^2 = 9(-1) + 9$
$27 + 9 = 27 + 9$	$-27 + 9 = -27 + 9$	$-1 + 1 = -9 + 9$
$36 = 36$	$-18 = -18$	$0 = 0$

Remember if both sides of the equation are equal, then the solution is considered to be correct for the replacement of the variable.

Pythagorean Theorem

Therorem -

A statement that is shown to be true by use of various axioms, definitions and other theorems in logical development.

Phythagorean Theorem

In any right triangle, the square of the length of the hypotenuse equals the sum of the squares of the lengths of the other two sides.

Figure 1

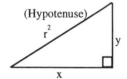

$r^2 = x^2 + y^2$
(Sometimes written as $c^2 = a^2 + b^2$)

By the use of this theorem, we can solve geometric problems consisting of right triangles with any two sides given. If we are given any two sides of a right triangle, we can use our ability to solve equations in order to find the third side.

Nate: The formula can be written as follows:

$x^2 + y^2 = r^2$ OR $a^2 + b^2 = c^2$

Let's take a typical example of a right triangle. We are going to find all three sides, thus checking ourselves as we go along.

Figure 2

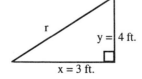

Now, using the formula $r^2 = x^2 + y^2$, we can solve for "r" (hypotenuse).

1. $r^2 = x^2 + y^2$

2. $r^2 = (3)^2 + (4)^2$ (Substitution)

3. $r^2 = 9 + 16$ Remember $(3)^2 = 3 \cdot 3 = 9$
                                          $(4)^2 = 4 \cdot 4 = 16$

4. $r^2 = 25$

Remember we are trying to find r **NOT** r^2. So we must get the $\sqrt{r^2}$ which gives us r. Since we are solving an equation, we must do the same to both sides.

5. $\sqrt{r^2} = \sqrt{25}$ (Find the square root of both sides)

6. $r = 5$ ft.

Now let's pretend that we do not know the length of side x. We will use the formula to find it. Let's draw the figure again:

Figure 3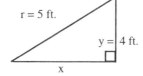

r = 5 ft.   y = 4 ft.   x

1. $r^2 = x^2 + y^2$

2. $(5)^2 = x^2 + (4)^2$ (Substitution)

3. $25 = x^2 + 16$ (Squaring 5 and 4)

4. $25 - 16 = x^2 + 16 - 16$ (Solving for x2)

5. $9 = x^2$ (Don't stop here)

6. $\sqrt{9} = \sqrt{x^2}$ (Square root)

7. 3 ft. $= x$ **OR** $x = 3$ ft.   (Answer)

Look at what we have just done. We knew that r = 3 earlier, but after finding r = 5, we pretended that we did not know the length of x and used the formula to find it. How about that? Did you like what we did? We hope that you are not getting bored with what we are doing, because we have only covered about one third of this book. We have two thirds left to go. We sincerely hope that you are getting some valuable learning from this experience. And, in the mean time, let's get busy, because 'it aint nothing to it, it's just the way you do it".

Now let's pretend that we do not know the length of side "y". We will also use the same formula to find it. Let's draw the figure one more time.

Figure 4.

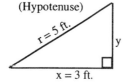

(Hypotenuse)
$r = 5$ ft.
y
$x = 3$ ft.

1. $r^2 = x^2 + y^2$

2. $(5)^2 = (3)^2 + y^2$ (Substitution)

3. $25 = 9 + y^2$ (Squaring 5 and 3)

4. $25 - 9 = 9 - 9 + 9^2$ (Solving for y^2)

5. $\sqrt{16} = \sqrt{y^2}$ (Don't stop here)

6. $16 = y^2$ (Square root)

7. 4 ft. $= y$ **OR** $y = 4$ ft.   (Answer)

Now let's check by using all three sides in the formula at one time.

$$x^2 + y^2 = r^2$$
$$(3)^2 + (4)^2 = (5)^2$$
$$9 + 16 = 25$$
$$25 = 25$$

Well, since we have finished studying through the formula, let's put it to work.

Let's Practice #28:

Use the following triangle in Exercises 1 - 6. Find the missing length correct to the nearest tenth. You may use a calculator if needed.

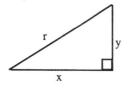

r
y
x

NOTE: $(\sqrt{a})^2 = a$, $(\sqrt{5})^2 = \sqrt{5} \cdot \sqrt{5} = \sqrt{25} = 5$

1) $y = 6$, $x = 8$, $r =$ _____ ? _____

2) $y = 5$, $x = 9$, $r =$ _____ ? _____

3) $y = 10$, $x = 24$, $r =$ _____ ? _____

4) $y =$ _____ ? _____, $x = \sqrt{10}$, $r = \sqrt{26}$

5) $y = 11$, $x =$ _____ ? _____, $r = 15$

6) $y = 12$, $x = 7$, $r =$ _____ ? _____

7) $y =$ _____ ? _____, $x = 5$, $r = 13$

8) $y = 5$, $x =$ _____ ? _____, $r = \sqrt{30}$

9) $y = \dfrac{\sqrt{3}}{4}$, $x = \dfrac{1}{4}$, $=$ _____ ? _____

10) $y = \dfrac{1}{5}$ $x = \dfrac{\sqrt{15}}{5}$ $r =$ _____ ? _____

If you had trouble with #4, #8, and #9, here are some look-alikes on the next page.

Look-A-Like #4: $y = \underline{\quad\quad}?\underline{\quad\quad}$, $x = \sqrt{12}$, $r = \sqrt{37}$

$$r^2 = x^2 + y^2$$

1. $(\sqrt{37})^2 = (\sqrt{12})^2 + y^2$ (Substitution)

2. $37 = 12 + y^2$ (Squaring $\sqrt{37}$ and $\sqrt{12}$)

3. $37 - 12 = 12 - 12 + y^2$ (Solving for y^2)

4. $25 = y^2$ (Don't stop here)

5. $\sqrt{25} = \sqrt{y^2}$

6. $5 = y$ **OR** $y = 5$ **(Answer)**

Look-A-Like #8: $y = 6$, $x = \underline{\quad\quad}?\underline{\quad\quad}$ $r = \sqrt{41}$

Remember: $(\sqrt{a})^2 = a$, $(\sqrt{3})^2 = \sqrt{3} \cdot \sqrt{3} = \sqrt{9} = 3$

1. $r^2 = x^2 + y^2$

2. $(\sqrt{41})^2 = x^2 + (6)^2$

3. $41 = x^2 + 36$

4. $41 - 36 = x^2 + 36 - 36$

5. $5 = x^2$

6. $\sqrt{5} = \sqrt{x^2}$

7. **Answer:** $2.2 = x$ **OR** $x = 2.2$

(Remember, to the nearest tenth.)

Look-A-Like #9: $y = \dfrac{\sqrt{5}}{6}$, $x = \dfrac{2}{6}$, $r = \underline{\quad\quad}?\underline{\quad\quad}$

1. $r^2 = x^2 + y^2$

2. $r^2 = \left(\dfrac{\sqrt{5}}{6}\right)^2 + \left(\dfrac{2}{6}\right)^2$ (Substitution)

3. $r^2 = \dfrac{5}{36} + \dfrac{4}{36}$ Squaring $\dfrac{\sqrt{5}}{6}$ and $\dfrac{2}{6}$

$\left(\textbf{NOTE:}\quad \left(\dfrac{2}{6}\right)^2 = \dfrac{2}{6} \cdot \dfrac{2}{6} = \dfrac{4}{36}\right)$

4.　　　$r^2 = \dfrac{9}{36}$

5.　　　$\sqrt{r^2} = \sqrt{\dfrac{9}{36}}$

6.　　　**Answer:** $r, = \dfrac{3}{6} = \dfrac{1}{2}$　**OR**　$r = \dfrac{1}{2}$

SIMPLIFYING ALGEBRAIC FRACTIONS

In order to simplify these types of fractions, we must use all of our skills for factoring polynomials. First, we must look at the numerator and the denominator to see if they can be factored. If so, then we will factor them both and apply the cancellation rule. In short, see what will divide out. The quotient is said to be in its simplest form, when the numerator and denominator have no common factor than 1 and -1.

Example #1:　　　　Write in the simplest form:

$$\dfrac{x^2 - 4,}{x^2 - 7x - 18}$$

1.　　$\dfrac{x^2 - 4}{x^2 - 7x - 18}$　　　　$= \dfrac{(x - 2)\,\cancel{(x + 2)}}{(x - 9)\,\cancel{(x + 2)}}$　　　(Fractor and cancel)

2.　　　　　　　　　　　$= \dfrac{x - 2}{x - 9}$　$(x \neq 9, x \neq -2)$　　**(Restrictions)** Answer

NOTE: The restrictions are based on the
　　　denominator before any cancellation is done.

Remember, you cannot divide by zero, so we must restrict the variable in the denominator so that the denominator does not equal zero., In Example #1, if x is equal to 9 or -2, the denominator will become zero and this would make the expression undefined. Thus violating the definition of a fraction. (A fraction is an expression in the form a/b, b ≠ 0.

Example #2:　　　　Write in the simplest form:
$$\dfrac{x^2 - 6x + 9}{x^2 - 9}$$

1.　　$\dfrac{x^2 - 6x + 9}{x^2 - 9}$　$= \dfrac{(x - 3)\,(x - 3)}{\cancel{(x - 3)}\,(x + 3)}$　　(Factor and cancel)

2.　　　　　$= \dfrac{(x - 3}{x + 3}$　　**(Restrictions)**
　　　　　　　　　$(x \neq 3, x \neq -3)$ Answer

Let's Practice #29:   (Note the restrictions for the variables.)

1) $\dfrac{3x - 6}{x - 2}$

2) $\dfrac{3x+y}{9x + 3y}$

3) $\dfrac{5x - 5y}{5x + 5y}$

4) $\dfrac{x^2 + 3x}{x^2 - 6x}$

5) $\dfrac{y^2 - xy}{y^2 + xy}$

6) $\dfrac{a}{3a^2 - a}$

7) $\dfrac{3a + 4}{9a^2 - 16}$

8) $\dfrac{4y + 20}{y^2 - 25}$

9) $\dfrac{2y^2 - 8}{y^2 + 4y - 12}$

10) $\dfrac{x^2 - 2x - 15}{x^2 + 10x + 21}$

11) $\dfrac{64x^2 - 4y^2}{y^2 + 3xy - 4x^2}$

12) $\dfrac{3a^2 - 3b^2}{3a(a - b) + 3b(a - b)}$

If you had trouble with #2 and #11, here are some look-alikes:

Look-A-Like #2: $\dfrac{2x + y}{6x + 3y} = \dfrac{\cancel{2x + y}}{3\cancel{(2x + y)}} = \dfrac{1}{3}$

$(x \neq -y, \ y \neq -2x)$

Let's look at how we find the restrictions. Take the denominator before we cancel and set it to be equal to zero, then solve.

$3(2x + y) = 0$ Remember solving equations in factored form.

$3(2x + y) = 0$

	Solve for x	Solve for y
$3 = 0$	$2x + y = 0$	$2x + y + 0$
(reject)	$2x + y - y = 0 - y$	$2x - 2x + y = 0 - 2x$
	$2x = -y$	$y = -2x$
	$\dfrac{2x}{2} = \dfrac{-y}{2}$	restriction
	$x = \dfrac{-y}{2}$	
	restriction	

62

Since restrictions are things we do not want to happen,

we do not want $x = \dfrac{-y}{2}$ and $y = -2x$. Thus, we write as our restrictions

$$\left(x \neq \dfrac{-y}{2}, \; y \neq -2x \right)$$

Look-A-Like #11: $\dfrac{36x^2 - 9y^2}{y^2 + xy - 2x^2}$

1. $\dfrac{36x^2 - 9y^2}{y^2 + xy - 2x^2} = \dfrac{9(4x^2 - y^2)}{(y + 2x)(y - x)}$ Remember the difference of squares.

$\qquad\qquad = \dfrac{9(2x - y)\,\cancel{(2x + y)}}{\cancel{(y + 2x)}\,(y - x)}$ **NOTE:** $2x + y = y + 2x$
                                                            Commutative axiom for addition.

$\qquad\qquad = \dfrac{9(2x - y)}{y - x}$ $(y \neq x, \; y \neq -2x)$
                                                            $(x \neq y, \; x \neq \dfrac{-y}{2})$

Again, let's look at how we find the restrictions. Look at the denominator before we cancel.

$(y + 2x)$	$(y - x) = 0$
Solve for x	Solve for y
$y + 2x = 0$	$y - x = 0$
$y - y + 2x = 0 - y$	$y - x + x = 0 + x$
$2x = -y$	$y = x$
$\dfrac{2x}{2} = \dfrac{-y}{2}$	
$x = \dfrac{-y}{2}$	
Solve for y	Solve for x
$y + 2, - 0$	$x = y$ (from above)
$y = 2x - 2x = 0 - 2x$	
$y = -2x$	

Restrictions: $(y \neq -2x, \; y \neq x, \; x \neq \dfrac{-y}{2}, \; x \neq y)$

MULTIPLYING ALGEBRAIC FRACTIONS

In order to multiply algebraic fractions, we must recall the basic rules for multiplying arithmetic fractions. When we multiply algebraic fractions, very much like arithmetic fractions, we must multiply the numerator of one fraction by the numerator of the other. This same process must also be done with the denominators.

Before we start, let's look at the the multiplication rule for fractions shown below. This rule permits us to express a product of two fractions as a quotient.

Multiplication Rule for Fractions

If a, b, x, and y are real numbers such that $b \neq 0$ and $y \neq 0$,

then: $\quad \dfrac{a}{b} \cdot \dfrac{x}{y} = \dfrac{ax}{by}$

Let's look at some examples:

Example #1: $\qquad \dfrac{3x^2}{2y} \cdot \dfrac{6y^2}{x} \qquad$ (Multiply)

1. $\quad \dfrac{3x^2}{2y} \cdot \dfrac{6y^2}{x} = \dfrac{3x^2 \cdot 6y^2}{2y \cdot x}$

2. $\quad \dfrac{9\cancel{x}\cancel{x}\cancel{y}}{\underset{2xy}{\cancel{18x^2y^2}}} = 9xy \qquad$ (Answer)
$\qquad\qquad 1$

Example #2: $\quad \dfrac{4a}{2b^2} \cdot \dfrac{3b^2}{a^2}$

1. $\quad \dfrac{4a}{2b^2} \cdot \dfrac{3b^2}{a^2} = \dfrac{4a \cdot 3b^2}{2b^2 \cdot a^2}$

2. $\quad = \dfrac{\overset{\cancel{6}\cdot\cancel{y}\cdot 1}{}}{\underset{\underset{a\,\cdot\,1}{2a^2b^2}}{\cancel{12ab^2}}} \qquad$ = Remember division of monomials

3. **Answer** $= \dfrac{6}{a}$

Example #3: $\dfrac{x-1}{2x-5} \cdot \dfrac{4x-10}{1-x}$ $\dfrac{-10}{}$

1. $\dfrac{x-1}{2x-5} \cdot \dfrac{4x-10}{1-x} = \dfrac{x-1}{\cancel{2x-5}} \cdot \dfrac{\overset{1}{2\cancel{(2x-5)}}}{-x+1}$

   (with 1 under $2x-5$)

NOTE: $-x+1 = -1(x-1)$ (factor out -1)

(Factor 4x - 10)

2. $\dfrac{2(\cancel{x-1})}{-1(\cancel{x-1})}$

3. **Answer:** $= -2$

Example #4: $\dfrac{y^2-3y-4}{y^2-2y} \cdot \dfrac{y^2-4}{y-4}$

1. $\dfrac{y^2-3y-4}{y^2-2y} \cdot \dfrac{y^2-4}{y-4} = \dfrac{\overset{1}{\cancel{(y-4)}}\,(y+1)}{y\cancel{(y-2)}} \cdot \dfrac{\overset{1}{\cancel{(y-2)}}(y+2)}{\cancel{y-4}}$ (factor)*

2. $= \dfrac{(y+1)(y+2)}{y}$ $(y \ne 0,\ y \ne 2,\ y \ne 4)$

Let's Practice #30:

Multiply each and write in the simplest form:

1) $\dfrac{32x^4}{y^3} \cdot \dfrac{y}{16x}$

2) $\dfrac{y+2}{y} \cdot \dfrac{y^2}{y^2-4}$

3) $\dfrac{x-1}{3} \cdot \dfrac{6}{x^2-1}$

4) $\dfrac{a+b}{a-b} \cdot \dfrac{a^2-b^2}{3a+3b}$

5) $\dfrac{x^2-y^2}{x} \cdot \dfrac{4}{2y+2x}$

6) $\dfrac{x^2-5x}{5} \cdot \dfrac{10x}{2x-10}$

7) $\dfrac{a^2+b^2}{a^2+2ab+b^2} \cdot \dfrac{3a^2-3b^2}{6}$

8) $\dfrac{x^2+5x+6}{4x-4} \cdot \dfrac{x^2-x}{x+2}$

9) $\dfrac{x^2+10x+21}{x^2-2x-15} \cdot \dfrac{x^2-8x+15}{x^2+5x-14}$

10) $\dfrac{2y3-5y2-3y}{y^2-4y} \cdot \dfrac{8-2y}{y^2-3y}$

11) $\dfrac{a^2 - 2a - 8}{a^2} \cdot \dfrac{a^2 + a - 2}{9a} \cdot \dfrac{54a^3}{a^4 - 5a^2 + 4}$

12) $\dfrac{y^3 + 3y^2 - 4y - 12}{2y^2 - 18} \cdot \dfrac{y^3 - 3y^2 + 3y - 9}{3y^3 - 12y}$

If you had trouble with #11 and 12, here are some look-alikes:

Look-A-Like #11: $\quad \dfrac{4z^2 - 4}{1 + z} \cdot \dfrac{1 + z}{2x} \cdot \dfrac{1 - 2z^2 + z^4}{2 + 2z}$

1. Factor: $\quad \dfrac{\overset{2}{\cancel{4}}(z^2 - 1)}{1 + z} \quad \dfrac{\overset{1}{\cancel{(1+z)}}}{\underset{1}{\cancel{2z}}} \quad \dfrac{(1 - z^2)(1 - z^2)}{\underset{1\quad 1}{2\cancel{(1+z)}}}$

2. Factor and simplify:

$$= \dfrac{(z - 1)\,\overset{1}{\cancel{(z+1)}}}{\underset{1}{\cancel{1+z}}} \cdot \dfrac{(1 - z^2)(1 - z^2)}{z}$$

Answer: $\quad = \dfrac{(z - 1)(1 - z^2)^2}{z}$

Look-A-Like #12: $\quad \dfrac{x^3 + 4x^2 - 4x - 16}{3(x^2 - 16)} \cdot \dfrac{x^3 - 4x^2 + 4x - 16}{3x^3 - 12x}$

$$= \dfrac{x^2(x + 4) - 4(x + 4)}{3(x - 4)(x + 4)} \cdot \dfrac{x^2(x - 4) + 4(x - 4)}{(3x(x^2 - 4)}$$

$$= \dfrac{\overset{1}{\cancel{(x^2 - 4)}}\,\overset{1}{\cancel{(x + 4)}}}{\underset{1\qquad 1}{3\cancel{(x - 4)}\,\cancel{(x + 4)}}} \cdot \dfrac{(x^2 + 4)\,\overset{1}{\cancel{(x - 4)}}}{\underset{1}{3x\cancel{(x^2 - 4)}}}$$

Answer: $\quad = \dfrac{x^2 + 4}{9x}$

DIVIDING ALGEBRAIC FRACTIONS

In order to divide algebraic fractions, we recall the basic rule for arithmetic fractions. The rule tells us to invert our divisor on the second step of the problem and multiply the fractions together:

1. $\frac{2}{3} \div \frac{1}{4}$ (Invert the divisor and multiply)

2. $\frac{2}{3}$ x $\frac{4}{1}$

3. $\frac{8}{3}$ or $2\frac{2}{3}$ (**Answer**)

Example #1: $\frac{3x^2y}{4} \div \frac{6y}{x}$

1. $\frac{3x^2y}{4} \div \frac{6y}{x}$

2. $\overset{1}{\cancel{3}}x^{\overset{2}{\cancel{2}}}\cancel{y} \cdot \dfrac{x}{\underset{2 \quad 1}{\cancel{6}\cancel{y}}}$ (Invert and multiply
                                   (Remember to cancel wherever you can)

3. $\dfrac{x^3}{8}$ (**Answer**)

In the next example, we will have to use our factoring skills to help control some numerators and denominators.

Example #2: $\dfrac{2x + 2y}{x^2} \div \dfrac{x^2 - y^2}{3x}$

1. $\dfrac{2x + 2y}{x^2} \div \dfrac{x^2 - y^2}{3x}$

2. $\dfrac{2\overset{1}{\cancel{(x+y)}}}{\underset{x}{\cancel{x^2}}} \cdot \dfrac{\overset{1}{\cancel{3x}}}{(x - y)\underset{1}{\cancel{(x+y)}}}$ (Invert, factor and multiply)

3. $\dfrac{6}{x(x - y)}$ (**Answer**)

Example #3: $\dfrac{6}{x^2 - 9} \div \dfrac{3x - 6}{x - 3} \cdot \dfrac{x - 2}{2}$

1. $\dfrac{6}{x^2 - 9} \div \dfrac{3x - 6}{x - 3} \cdot \dfrac{x - 2}{2}$

2. $\dfrac{\overset{1}{\overset{\cancel{2}}{\cancel{6}}}}{(\cancel{x - 3})(x + 3)} \cdot \dfrac{\overset{1}{\cancel{x - 3}}}{\underset{1}{\cancel{3(x - 2)}}} \cdot \dfrac{\overset{1}{\cancel{x - 2}}}{\underset{1}{\cancel{2}}}$ (Invert, factor and multiply)

3. $\dfrac{1}{x + 3}$ **(Answer)**

Let's practice #31: Divide. Write each in its simplest form.

1) $\dfrac{a + b}{6} \div \dfrac{a + b}{12}$

2) $\dfrac{a - 1}{3} \div \dfrac{a^2 - 1}{6}$

3) $\dfrac{x^2 - 4}{2x} \div (x - 2)$

4) $\dfrac{a - b}{a + b} \div \dfrac{1}{(a + b)^2}$

5) $\dfrac{x^2 + 9x + 14}{x^2 - 9} \div \dfrac{x^2 + 5x - 14}{x^2 - 5x + 6}$

6) $\dfrac{a^2 - 36}{6a - 36} \div \dfrac{a^2 + 12a + 36}{a^3 + 6a^2}$

7) $\dfrac{6y}{6y - 14} \div \dfrac{21}{9y - 21}$

8) $\dfrac{a^4 - b^4}{4a - 4b} \div \dfrac{a^2 + b^2}{8}$

9) $\dfrac{3x^2 - 9x + 6}{3 - 3x} \div \dfrac{2x^2 - 10x + 12}{6 - 2x}$

10) $\dfrac{3a^2 - 14a + 8}{6 - 25a + 24a^2} \div \dfrac{2a^2 - 3a - 20}{16a^2 + 34a - 15}$

HINT: $6 - 25a + 24a^2$ can be written in standard form as $24a^2 - 25a + 6$.

11) $\dfrac{x^2 - 4x + 4}{x^2 - 2x} \div \dfrac{10 - 7x + x^2}{2x^2 - 13x + 15}$

12) $\dfrac{a^2}{a^2 - b^2} \cdot \dfrac{a - b}{a + b} \div \dfrac{a}{(a + b)^2}$

If you had trouble with #9 and #12, here are some look-alikes:

Look-A-Like #9: $\qquad \dfrac{2x^2 - 8x + 6}{2 - 2x} \div \dfrac{3x^2 - 21x + 36}{12 - 3x}$

1. $\quad = \quad \dfrac{2x^2 - 8x + 6}{2 - 2x} \cdot \dfrac{12 - 3x}{3x^2 - 21x + 36}$

2. $\quad = \quad \dfrac{2(x^2 - 4x + 3)}{-2x + 2} \cdot \dfrac{-3x + 12}{3(x^2 - 7x + 12)}$

   Remember: $-3x + 12$ and $-2x + 2$ are the standard forms

3. $\quad = \quad \dfrac{2\overset{1}{\cancel{(x - 3)}}\,\overset{1}{\cancel{(x - 1)}}}{\underset{1}{-2\cancel{(x - 1)}}} \cdot \dfrac{\overset{1}{-3\cancel{(x - 4)}}}{\underset{1}{3}\underset{1}{\cancel{(x - 4)}\cancel{(x - 3)}}}$

   Factor out -3 also -2.

4. $\quad = \quad \dfrac{2}{-2} \cdot \dfrac{-3}{3}$

   Remember: $- \cdot - = +$

5. $\quad = \quad \dfrac{6}{6}$

6. $\quad = \quad 1$

   (Answer)

For look-alike #12, see Example #3

ADDING AND SUBTRACTING ALGEBRAIC FRACTIONS

To add fractions with like denominators, we add their numerators. To subtract fractions with like denominators, we subtract their numerators.

Example #1: $\qquad \dfrac{2x}{z + 1} + \dfrac{y}{z + 1} = \dfrac{2x + y}{z + 1}$ **Answer**

Example #2: $\qquad \dfrac{4}{x - y} + \dfrac{2}{y - x} = \dfrac{4}{x - y} + \dfrac{2}{-x + y}$

NOTE: Factor out -1 and divide -1 into the numerator "2". $\qquad = \dfrac{4}{x - y} + \dfrac{2}{-1(x - y)}$

Now we have:

$$= \frac{4}{x - y} + \frac{-2}{x - y}$$

$$= \frac{2}{x - y} \qquad \textbf{Answer}$$

Example #3:

$$\frac{x^2}{(x + 3)^2} - \frac{6x - 5}{(x + 3)^2} = \frac{x^2 - (6x - 5)}{(x + 3)^2}$$

$$= \frac{x^2 - 6x + 5}{(x + 3)^2}$$

* Remember when there is a negative sign between the fractions, you must multiply the negative sign times every term in the numerator that follows:

Since addition and subtraction of fractions with <u>unlike</u> denominators is usually the topic which gives most students trouble, we will want to focus our attention to it at this time.

To add or subtract fractions with <u>unlike</u> denominators, we must first express the fractions with a common denominator. The results should always be written in the simplest form.

Before we start, let's look at a valuable tool that we must learn to use. It is called an identity element or, sometimes, a unit element.

Identity Element - A mathematical expression that is equal to 1 as it relates to a fraction.

Example: $\dfrac{2}{2}$, $\dfrac{3}{3}$, $\dfrac{a}{a}$, $\dfrac{x}{x}$, $\dfrac{x + 1}{x + 1}$, etc.

Let's see how, this valuable tool can be used.

Example #1: $\dfrac{3}{y} + \dfrac{2}{y^2}$

1. First look at the denominator to see what can be done to make them alike. If they were both y^2, we would have it made. Since the fraction on the right has y^2 as it's denominator, let's operate on the one on the left side.

2. $\dfrac{3}{y} \left(\right) + \dfrac{2}{y^2}$ What will the identity element be?
Remember we can multiply "1" times any term
and not change the value of that term.

3. $\dfrac{3}{y} \left(\dfrac{y}{y} \right) + \dfrac{2}{y^2}$ Multiply by $\dfrac{y}{y}$

4. $\dfrac{3y}{y^2} + \dfrac{2}{y^2}$ Add like denominators

5. $\dfrac{3y + 2}{y^2}$ Answer

Example #2: $\dfrac{4}{7a} + \dfrac{3}{ab}$

Since there is nothing that we can multiply times one denominator to make it look like the other, we will have to operate on both fractions. Now, if we could have 7ab in both denominators, we would be on time. Let's try something:

1. $\dfrac{4}{7a} \left(\right) + \dfrac{3}{ab} \left(\right)$ What elements will we need?

2. $\dfrac{4}{7a} \left(\dfrac{b}{b} \right) + \dfrac{3}{ab} \left(\dfrac{7}{7} \right)$ Multiply by $\dfrac{b}{b}$ and $\dfrac{7}{7}$

3. $\dfrac{4b}{7ab} + \dfrac{21}{7ab}$ Add

4. $\dfrac{4b + 21}{7ab}$ **Answer**

Example #3: $\dfrac{3}{x - 2} + \dfrac{1}{x + 2}$

      Since we can not multiply anything times either denominator to make one look like the other, we must operate on both:

1. $\dfrac{3}{x - 2} + \dfrac{1}{x + 2}$

2. $\dfrac{3}{x - 2} \left(\dfrac{x + 2}{x + 2} \right) + \dfrac{1}{x + 2} \left(\dfrac{x - 2}{x - 2} \right)$

When this happens, we must key of f of each denominator. We must take each denominator, make an identity element from it and operate on the opposite fraction:

3. $\dfrac{3x + 6}{(x - 2)(x + 2)}$ + $\dfrac{x - 2}{(x - 2)(x + 2)}$ Add

4. $\dfrac{3x + 6 + x - 2}{(x - 2)(x + 2)}$ Add like terms

5. $\dfrac{4x + 4}{(x - 2)(x + 2)}$ **Answer**

Example #4: $\dfrac{a + 1}{a} + \dfrac{a + 2}{a + 1}$

1. $=$ $\dfrac{a + 1}{a}\left(\dfrac{a + 1}{a + 1}\right) - \dfrac{a + 2}{a + 1}\left(\dfrac{a}{a}\right)$

2. $=$ $\dfrac{a^2 + 2a + 1}{a(a + 1)} - \dfrac{a^2 + 2a}{a(a + 1)}$

3. $=$ $\dfrac{a^2 + 2a + 1 - (a^2 + 2a)}{a(a + 1)}$ Remember to multiply the negative sign times the entire numerator.

4. $=$ $\dfrac{a^2 + 2a + 1 - a^2 - 2a}{a(a - 1)}$

5. $=$ $\dfrac{1}{a(a + 1)}$ **Answer**

Let's Practice #32:   Simplify:

1) $\dfrac{a}{a - 1} - \dfrac{3a}{a^2 - 1}$ Remember $a^2 - 1 = (a - 1)(a + 1)$

2) $\dfrac{y + 1}{y} - \dfrac{y}{y + 1}$ 3) $\dfrac{1}{6x - 6} - \dfrac{1}{3x^2 - 3}$

4) $\dfrac{4}{x + 5} + \dfrac{3}{x - 5}$ 5) $\dfrac{y}{x + y} + \dfrac{x}{x - y}$

6) $\dfrac{2}{a^2 - 2a} - \dfrac{3}{a^2 - a - 2}$ 7) $\dfrac{y}{x^2 - xy} - \dfrac{x}{xy - x^2}$

8) $\dfrac{1}{x^2 - 1} + \dfrac{2}{x^2 + 2x + 1}$ 9) $\dfrac{2}{a^2 - a - 2} + \dfrac{1}{a^2 - 4}$

10) $\dfrac{3a}{a^3 - 5a^2} + \dfrac{5}{a^2 + 5a}$

11) $\dfrac{2x}{x^2 + 3x + 2} - \dfrac{3x - 6}{x^2 + 4x + 4}$

12) $\dfrac{y^2 + 1}{y^2 - 1} + \dfrac{1}{y + 1} + \dfrac{1}{y - 1}$

If you had trouble with #10, #11, and #12, here are some look-alikes:

Look-A-Like #10: $\dfrac{2a}{a^3 - 3a^2} + \dfrac{3}{a^2 + 3a}$

1. $\dfrac{2a}{a^2(a - 3)} + \dfrac{3}{a(a + 3)}$ Factor the denominators.

What elements will we need?

2. $\dfrac{2a}{a^2(a - 3)}\left(\dfrac{a + 3}{a + 3}\right) + \dfrac{3}{a(a + 3)}\left(\dfrac{a(a - 3)}{a(a - 3)}\right)$ Multiply

3. $\dfrac{2a^2 + 6a}{a^2(a - 3)(a + 3)} + \dfrac{3a^2 - 9a}{a^2(a - 3)(a + 3)}$

Remember $(a + 3)(a - 3) = (a - 3)(a + 3)$ Commutative axiom for multiplication.

4. $\dfrac{2a^2 + 6a + 3a^2 - 9a}{a^2(a - 3)(a + 3)}$ Add like term

5. $\dfrac{5a^2 - 3a}{a^2(a - 3)(a + 3)}$ Factor the numerator

6. $\dfrac{\overset{1}{\cancel{a}}(5a - 3)}{\underset{a}{\cancel{a^2}}(a - 3)(a + 3)}$

7. $\dfrac{5a - 3}{a(a - 3)(a + 3)}$ **Answer**

Look-A-Like #11: $\dfrac{3x}{x^2 - 3x + 2} - \dfrac{3x + 6}{x^2 - 4x + 4}$

1. $\dfrac{3x}{(x - 2)(x - 1)} - \dfrac{3x + 6}{(x - 2)(x - 2)}$ Factor the denominators

2. $$\frac{3x}{(x-2)(x-1)}\left(\frac{x-2}{x-2}\right) \quad - \quad \frac{3x+6}{(x-2)(x-2)}\left(\frac{x-1}{x-1}\right)$$

3. $$\frac{3x^2-6x}{(x-2)(x-2)(x-1)} - \frac{3x^2+3x-6}{(x-2)(x-2)(x-1)}$$

4. $$\frac{3x^2-6x-(3x^2+3x-6)}{(x-2)(x-2)(x-1)}$$ Subtract or move the parenthesis by multiplying through with -1

5. $$\frac{3x^2-6x-3x^2-3x+6}{(x-2)(x-2)(x-1)}$$

6. $$\frac{-9x+6}{(x-2)(x-2)(x-1)}$$ Factor the numerator

7. $$\frac{-3(3x-2)}{(x-1)(x-2)^2}$$ **Answer**

Remember $(x-2)(x-2) = (x-2)^2$

Look-A-Like #12: $$\frac{y}{2y-1} \quad + \quad \frac{y-1}{2y+1} \quad + \quad \frac{2y}{4y^2-1}$$

1. $$\frac{y}{2y-1} \quad + \quad \frac{y-1}{2y+1} \quad + \quad \frac{2y}{4y^2-1}$$

2. $$\frac{y}{2y-1} \quad + \quad \frac{y-1}{2y+1} \quad + \quad \frac{2y}{(2y-1)(2y+1)}$$ Look at the denominator

3. $$\frac{y}{2y-1}\left(\frac{2y+1}{2y+1}\right) \quad + \quad \frac{y-1}{2y+1}\left(\frac{2y-1}{2y-1}\right) \quad + \quad \frac{2y}{(2y-1)(2y+1)}$$

4. $$\frac{2y^2+y}{(2y-1)(2y+1)} \quad + \quad \frac{2y^2-y-1}{(2y-1)(2y+1)} \quad + \quad \frac{2y}{(2y-1)(2y+1)}$$

5. $$\frac{2y^2+y+2y^2-y-1+2y}{(2y-1)(2y+1)}$$ Add like terms

6. $$\frac{4y^2+2y-1}{(2y-1)(2y+1)}$$ **Answer**

PROPORTIONS

An equation that expresses two ratios being equal is called a proportion. We can write proportions in several different ways:

$1 : 2 = 2 : 4$ Read: "1 is to 2 as 2 is to 4."

$\dfrac{x}{8} = \dfrac{3}{4}$ Read: " x divided by 8 equals 3 divided by 4."

$$\dfrac{w}{x} = \dfrac{y}{z}$$

We can use the multiplication property of equality to show that in a proportion the product of the extremes equals the product of the means.

$$\dfrac{w}{\cancel{x}}(\cancel{x}z) = \dfrac{y}{\cancel{z}}(x\cancel{z})$$

$$wz = yx$$

Example #1:   Solve the proportion $\dfrac{4}{x} = \dfrac{2}{9}$

              $\dfrac{4}{x} \diagdown\!\!\!\!\diagup \dfrac{2}{9}$ Use the cross product method.

              $4 \cdot 9 = 2 \cdot x$

              $36 = 2x$

              $\dfrac{2x}{2} = \dfrac{36}{2}$

Answer: $x = 18$

Example #2:   Solve the proportion $\dfrac{2}{3} = \dfrac{16}{x}$

1. $\dfrac{2}{3} \diagdown\!\!\!\!\diagup \dfrac{16}{x}$

2. $2 \cdot x = 16 \cdot 3$

3. $2x = 48$

4. $\dfrac{2x}{2} = \dfrac{48}{2}$

5.   **Answer:** $x = 24$

Example #3: Solve the proportion $\dfrac{x}{7} = \dfrac{3}{5}$

1. $\dfrac{x}{7} \times \dfrac{3}{5}$ =

2. $5 \cdot x = 3 \cdot 7$

3. $5x = 21$

4. $\dfrac{5x}{5} \quad \dfrac{21}{5}$

5. **Answer:** $x = \dfrac{21}{5}$

Let's Practice #33:

1) $\dfrac{y}{8} = \dfrac{3}{6}$ 2) $\dfrac{y}{9} = \dfrac{7}{3}$ 3) $\dfrac{3}{5} = \dfrac{6y}{10}$

4) $\dfrac{4}{5} = \dfrac{3x}{7}$ 5) $\dfrac{3}{8x} = \dfrac{4}{5}$ 6) $\dfrac{2}{9} = \dfrac{3}{y}$

Before we do anymore, let's look at another type of proportional expression:

Example #4:   Solve $\dfrac{x-4}{4} = \dfrac{x}{6}$

1. $\dfrac{x-4}{4} \times \dfrac{x}{6}$ Cross product

2. $6(x-4) = x(4)$ Use the distributive axiom

3. $6x - 24 = 4x$ Now solve for x

4. $6x - 24 + 24 = 4x + 24$

5. $6x = 4x + 24$

6. $6x - 4x = 4x - 4x + 24$

76

7. $2x = 24$

8. $\dfrac{2x}{2} = \dfrac{24}{2}$

9.   **Answer** $x = 12$

Let's Continue Practice #33:   Remember the lesson on solving equations

7) $\dfrac{5}{x-2} = \dfrac{15}{x+4}$ 8) $\dfrac{2y-1}{6} = \dfrac{4y-3}{7}$

9) $\dfrac{a}{3} = \dfrac{2a-1}{5}$ 10) $\dfrac{x-3}{8} = \dfrac{x}{5}$

11) $\dfrac{2x+2}{16} = \dfrac{3x+1}{3}$ 12) $\dfrac{a+7}{2} = 3a$

Equations with Fractional Coefficients

We can solve equations with fractional coefficients by using the least common denominator of all the fractions showing in the equation. We can transform the equation by multiplying both sides by the least common denominator. Then, we proceed to solve the transformed equation.

NOTE: LCD means Least Common Denominator

Example #1:   Solve: $\dfrac{3x-2}{2} + \dfrac{x}{5} = 16$

1. $\dfrac{3x-2}{2} + \dfrac{x}{5} = 16$

2. $10\left[\dfrac{3x-2}{2} + \dfrac{x}{5}\right] = 10[16]$ Multiply by LCD

3. $\dfrac{\overset{5}{\cancel{10}}}{1} \cdot \dfrac{3x-2}{\underset{1}{\cancel{2}}} + \dfrac{\overset{2}{\cancel{10}}}{1}\left(\dfrac{x}{\underset{1}{\cancel{5}}}\right) = 10[16]$

4. $5(3x-2) + 2x = 160$

5. $15x - 10 + 2x = 160$

6. $17x + 10 = 160$

7. $17x - 10 + 10 = 160 + 10$

8. $17x = 170$

9. $\dfrac{17x}{17} = \dfrac{170}{17}$

10.   **Answer:** $x = 10$

Example #2:   Solve: $\dfrac{x+1}{5} - \dfrac{3}{2} = \dfrac{3x-6}{10}$

1. $\dfrac{x+1}{5} - \dfrac{3}{2} = \dfrac{3x-6}{10}$

2. $10\left[\dfrac{x+1}{5} - \dfrac{3}{2}\right] = 10\left[\dfrac{3x-6}{10}\right]$

3. $\dfrac{\overset{2}{\cancel{10}}}{1} \cdot \dfrac{x+1}{\underset{1}{\cancel{5}}} - \dfrac{\overset{5}{\cancel{10}}}{1} \cdot \dfrac{3}{\underset{1}{\cancel{2}}} = \dfrac{\overset{1}{\cancel{10}}}{1} \cdot \dfrac{3x-6}{\underset{1}{\cancel{10}}}$

4. $2x + 2 - 15 = 3x - 6$

5. $2x - 13 = 3x - 6$

6. $2x - 13 + 13 = 3x - 6 + 13$

7. $2x = 3x + 7$

8. $2x - 3x = 3x - 3x + 7$

9. $-x = 7$

10. $(-1)(-x) = 7(-1)$

11.   **Answer:** $x = -7$

Example #3:   Solve: $\dfrac{5}{7}x - \dfrac{1}{21}x + 2 = 0$

1. $\dfrac{5}{7}x - \dfrac{1}{21}x + 2 = 0$

2. $21\left[\dfrac{5}{7}x - \dfrac{1}{21}x + 2\right] = 21(0)$

3. $\dfrac{\overset{3}{\cancel{21}}}{1} \cdot \dfrac{5}{\underset{1}{\cancel{7}}}x - \dfrac{\overset{1}{\cancel{21}}}{1} \cdot \dfrac{1}{\underset{1}{\cancel{21}}}x + \dfrac{21}{1} \cdot \dfrac{2}{1} = 0$

4. $15x - x + 42 = 0$

5. $14x + 42 = 0$

6. $14x + 42 - 42 = 0 - 42$

7. $14x = -42$

8. $\dfrac{14x}{14} = \dfrac{-42}{14}$

9. $x = -3$

Let's Practice #34:

1) $\dfrac{x}{3} + \dfrac{2x}{5} = 11$

2) $\dfrac{3}{8}x - \dfrac{1}{4}x = 2$

3) $\dfrac{4}{5}x - \dfrac{1}{15}x + 1 = 0$

4) $\dfrac{a+8}{16} - \dfrac{a-4}{12} = 1$

5) $\dfrac{a-5}{8} - \dfrac{2a+6}{9} + 1 = 0$

6) $\dfrac{x-3}{5} - \dfrac{x+2}{15} - \dfrac{2}{3} = 0$

7) $\dfrac{4x+1}{3} - \dfrac{2x+1}{5} = \dfrac{3}{5}$

8) $\dfrac{y+3}{3} - \dfrac{y}{4} = \dfrac{y-2}{5}$

9) $\dfrac{1}{2}(y-5) - (y+1) = \dfrac{1}{4}(y-12)$

10) $\dfrac{5y-1}{2} - \dfrac{3y+1}{4} = \dfrac{9}{2}$

11) $\dfrac{3a-1}{4} - \dfrac{9-a}{6} = \dfrac{14}{3}$

12) $\dfrac{a+8}{16} + \dfrac{a-4}{12} = 1$

If you had trouble with #9 and #11, here are some look-alikes:

Look-A-Like #9: $\dfrac{1}{3}(y-3) - (y+2) = \dfrac{1}{2}(y-8)$

1. Simplify: $\dfrac{1}{3} \cdot \dfrac{y-3}{1} - y + 2 = \dfrac{1}{2} \cdot \dfrac{y-8}{1}$

2. $\dfrac{y-3}{3} - y + 2 = \dfrac{y-8}{2}$

3. Multiply by LCD: $\dfrac{6}{1}\left[\dfrac{y-3}{3} - \dfrac{y+2}{1}\right] = \dfrac{6}{1}\left[\dfrac{y-8}{2}\right]$

4. $\dfrac{\overset{2}{\cancel{6}}}{1} \cdot \dfrac{y-3}{\underset{1}{\cancel{3}}} - \dfrac{6}{1} \cdot \dfrac{y+2}{1} = \dfrac{\overset{3}{\cancel{6}}}{1} \cdot \dfrac{y-8}{\underset{1}{\cancel{2}}}$

5. Multiply $\quad 2(y-3) - 6(y+2) = 3(y-8)$

6. Solve for y: $\quad \underline{2y} - 6 - \underline{6y} - 12 = 3y - 24 \qquad$ Remember: $- + - = -$

7. $\qquad -4y - 18 = 3y - 24$
   Remember: Solving equations with a variable on both sides.

8. $\qquad -4y - 18 + 18 = 3y - 24 + 18$

9. $\qquad -4y = 3y - 6$

10. $\qquad -4y - 3y = -6$

11. $\qquad -7y = -6$

12. $\qquad \dfrac{-7y}{-7} = \dfrac{-6}{-7}$

13. **Answer:** $\quad y = \dfrac{6}{7}$

Let's Check

1. $\dfrac{1}{3}(y-3) - (y+2) = \dfrac{1}{2}(y-8)$

2. $\dfrac{1}{3}\left(\dfrac{6}{7} - 3\right) - \left(\dfrac{6}{7} + 2\right) = \dfrac{1}{2}\left(\dfrac{6}{7} - 8\right)$

Remember the identity element.

3. $\dfrac{1}{3}\left(\dfrac{6}{7} - \dfrac{3}{1} \cdot \dfrac{7}{7}\right) - \left(\dfrac{6}{7} + \dfrac{2}{1} \cdot \dfrac{7}{7}\right) = \dfrac{1}{2}\left(\dfrac{6}{7} - \dfrac{8}{1} \cdot \dfrac{7}{7}\right)$

4. $\dfrac{1}{3}\left(\dfrac{6}{7} - \dfrac{21}{7}\right) - \left(\dfrac{6}{7} - \dfrac{14}{7}\right) = \dfrac{1}{2}\left(\dfrac{6}{7} - \dfrac{56}{7}\right)$

5. $\quad \dfrac{1}{\cancel{3}}_{1}\left(\dfrac{-\cancel{15}^{5}}{7}\right)-\left(\dfrac{20}{7}\right)=\dfrac{1}{\cancel{2}}_{1}\left(\dfrac{-\cancel{50}^{25}}{7}\right)$

6. $\quad \dfrac{-5}{7}-\dfrac{20}{7}=\dfrac{-25}{7}$

7. **Answer:** $\quad \dfrac{-25}{7}=\dfrac{-25}{7}$

HOW ABOUT THAT! IT CHECKS OUT CORRECTLY!

Look-A-Like #11: $\qquad \dfrac{2a+1}{3}-\dfrac{9-a}{4}=\dfrac{9}{2}$

1. Multiply by LCD: $\qquad \dfrac{12}{1}\left[\dfrac{2a+1}{3}-\dfrac{9-a}{4}\right]=\dfrac{12}{1}\left[\dfrac{9}{2}\right]$

2. $\quad \dfrac{\cancel{12}^{4}}{1}\cdot\dfrac{2a+1}{\cancel{3}_{1}}-\dfrac{\cancel{12}^{3}}{1}\cdot\dfrac{9-a}{\cancel{4}_{1}}=\dfrac{\cancel{12}^{6}}{1}\cdot\dfrac{9}{\cancel{2}_{1}}$

3. $\quad 4(2a+1)-3(9-a)=6(9)$
   Remember $-\bullet-=+$

   $\quad 8a+4+27+3a=54$

4. Combine like terms: $11a-23=54$

5. Solve for a: $\quad 11a-23+23=54+23$

   $\qquad \dfrac{11a}{11}=\dfrac{77}{11}$

6. **Answer:** $\quad a=7$

   Let's Check

   $\qquad \dfrac{2a+1}{3}-\dfrac{9-a}{4}=\dfrac{9}{2}$

   $\qquad \dfrac{2(7)+1}{3}-\dfrac{9-7}{4}=\dfrac{9}{2}$

$$\frac{14+1}{3} - \frac{2}{4} = \frac{9}{2}$$

$$\frac{15}{3} - \frac{2}{4} = \frac{9}{2}$$

Remember the identity element.

$$\frac{15}{3}\left(\frac{4}{4}\right) - \frac{2}{4}\left(\frac{3}{3}\right) = \frac{9}{2}$$

$$\frac{60}{12} - \frac{6}{12} = \frac{9}{2}$$

$$\frac{54}{12} = \frac{9}{2} \qquad \text{Reduce} \qquad \left(\frac{54}{12} = \frac{9}{2}\right)$$

$$\frac{9}{2} = \frac{9}{2}$$

HOW ABOUT THAT! It checks also.

An equation that has a variable in the denominator of one or more terms is called a **fractional equation.** We can use the identity element and the cross product method to solve fractional equations.

Example #1: Solve: $\dfrac{a+1}{2a-2} = \dfrac{a}{6} + \dfrac{1}{a-1}$ $\quad (a \neq 1)$

Remember the definition for a fraction.

1. Factor the denominator on the left side:
$$\frac{a+1}{2(a-1)} = \frac{a}{6} + \frac{1}{a-1}$$

2. Now, looking on our right, we can multiply:
$$\frac{1}{a-1} \left(\frac{2}{2}\right) \text{ and get } \frac{2}{2(a-1)}$$

which has a common denominator with: $\dfrac{a+1}{2(a-1)}$

Let's try something: $\dfrac{a+1}{2(a-1)} = \dfrac{a}{6} + \dfrac{1}{a-1}\left(\dfrac{2}{2}\right)$

$$\frac{a+1}{2(a-1)} = \frac{a}{6} + \frac{2}{2(a-1)}$$

Now let's subtract $\dfrac{2}{2(a-1)}$ from both sides:

We want to get one fraction on each side in order to do a cross product.

3. $\dfrac{a+1}{2(a-1)} - \dfrac{2}{2(a-1)} = \dfrac{a}{6} + \dfrac{2}{2(a-1)} - \dfrac{2}{2(a-1)}$

Yes, this is what we want.

4. Combine the terms in the numerator: $\dfrac{a+1-2}{2(a-1)} = \dfrac{a}{6}$

5. $\dfrac{a-1}{2(a-1)} \diagup\!\!\!\!\diagdown \dfrac{a}{6}$ $\quad$ NOTE: $a \cdot 2(a-1) = 2a(a-1)$

6. $6(a-1) = 2a(a-1)$

7. Set to equal zero: $6a - 6 = 2a^2 - 2a$

8. Combine like terms: $6a - 6a - 6 = 2a^2 - 2a - 6a$

9. $-6 = 2a^2 - 8a$

10. $-6 + 6 = 2a^2 - 8a + 6$

11. $0 = 2a^2 - 8a + 6$

$$0 = (2a - 2)(a - 3)$$

$2a - 2 = 0$	$a - 3 = 0$
$2a - 2 + 2 = 0 + 2$	$a - 3 + 3 = 0 + 3$
$2a = 2$	$a = 3$ (Answer)
$\dfrac{2a}{2} = \dfrac{2}{2}$	
(Reject) $a = 1$	

Let's Check a = 3

$$\frac{a + 1}{2a - 2} = \frac{a}{6} + \frac{1}{a - 1}$$

$$\frac{3 + 1}{2(3) - 2} = \frac{3}{6} + \frac{1}{3 - 1}$$

$$\frac{4}{6 - 2} = \frac{3}{6} + \frac{1}{2}$$

$$\frac{4}{4} = \frac{1}{2} + \frac{1}{2}$$

$$1 = 1$$

Example #2: Solve $\dfrac{3x + 5}{6} - \dfrac{10}{x} = \dfrac{x}{2}$

1. $\dfrac{3x + 5}{6} - \dfrac{10}{x} = \dfrac{x}{2}\left(\dfrac{3}{3}\right)$

2. $\dfrac{3x + 5}{6} - \dfrac{10}{x} = \dfrac{3x}{6}$ Move $\dfrac{3x + 5}{6}$ to the right side

3. $\dfrac{3x + 5}{6} - \dfrac{3x + 5}{6} - \dfrac{10}{x} = \dfrac{3x}{6} - \dfrac{3x + 5}{6}$

4. $\dfrac{-10}{x} = \dfrac{3x - (3x + 5)}{6}$

5. Cross product: $\dfrac{-10}{x} \diagup\!\!\!\diagdown \dfrac{3x - 3x - 5}{6}$

6. $6(-10) = x(-5)$

7. $-60 = -5x$

8. $\dfrac{-60}{-5} = \dfrac{-5x}{-5}$

9. **Answer:** $12 = x$ or $x = 12$

Let's Check

$$\dfrac{3x + 5}{6} - \dfrac{10}{x} = \dfrac{x}{2}$$

$$\dfrac{3(12) + 5}{6} - \dfrac{10}{12} = \dfrac{12}{2}$$

$$\dfrac{36 + 5}{6} - \dfrac{10}{12} = 6$$

$$\dfrac{41}{6} - \dfrac{10}{12} = 6$$

$$\dfrac{41}{6}\left(\dfrac{2}{2}\right) - \dfrac{10}{12} = 6$$

$$\dfrac{82}{12} - \dfrac{10}{12} = 6$$

$$\dfrac{72}{12} = 6$$

$$6 = 6$$

Example #3: Solve: $\dfrac{1}{y - 4} = \dfrac{2}{y^2 - 16}$ $(y \neq 4)$

1. Cross product: $\dfrac{1}{y - 4} \diagup\!\!\!\diagdown \dfrac{2}{y^2 - 16}$

2. $1(y^2 - 16) = 2(y - 4)$

3. $y^2 - 16 = 2y - 8$

4. $y^2 - 16 - 2y = 2y - 2y - 8$

5. $y^2 - 16 - 2y = -8$

6. $y^2 - \underline{16} - 2y + \underline{8} = -8 + 8$

7. $y^2 - 2y - 8 = 0$

8. $(y - 4)(y + 2) = 0$

$y - 4 = 0$	$y + 2 = 0$
$y - 4 + 4 = 0 + 4$	$y + 2 - 2 = 0 - 2$
Reject $y = 4$	$y = -2$

The check is left for you to do.

Let's look at another way to solve fractional equations. This time we will use the multiplication property of equality to solve this type of equation. We will need the LCD.

Example #4:   Solve $\dfrac{3x}{x - 1} - \dfrac{4}{x + 1} = \dfrac{4}{x^2 - 1}$ $(x \neq 1, -1)$

1.   Factor $\dfrac{3x}{x - 1} - \dfrac{4}{x + 1} = \dfrac{4}{(x - 1)(x + 1)}$

2.   Multiply by the LCD which is $(x - 1)(x + 1)$

3. $\left[\dfrac{(x - 1)(x + 1)}{1}\right]\left[\dfrac{3x}{x - 1} - \dfrac{4}{x + 1}\right] = \left[\dfrac{(x - 1)(x + 1)}{1}\right]\left[\dfrac{4}{(x - 1)(x + 1)}\right]$

4. $\dfrac{(x - 1)(x + 1)}{1} \cdot \dfrac{3x}{x - 1} - \dfrac{(x - 1)(x + 1)}{1} \cdot \dfrac{4}{x + 1} = \dfrac{(x - 1)(x + 1)}{1} \cdot \dfrac{4}{(x - 1)(x + 1)}$

5. $(x + 1)\,3x - (x - 1)\,4 = 4$

6. $3x^2 + 3x - (4x - 4) = 4$

86

7. $3x^2 + 3x - 4x + 4 = 4$

8. $3x^2 - x + 4 = 4$

9. $3x^2 - x + 4 - 4 = 4 - 4$

10. $3x^2 - x = 0$

11. $x(3x - 1) = 0$

$$x = 0 \qquad \Big| \qquad 3x - 1 - 0$$

$$3x - 1 + 1 = 0 + 1$$

$$3x = 1$$

$$\frac{3x}{3} = \frac{1}{3}$$

$$x = \frac{1}{3}$$

Answer: $0, \dfrac{1}{3}$ **NOTE:** Sometimes we get two solutions
You may want to check them <u>both</u>.

Let's Practice #35:

1) $\dfrac{3}{a} - \dfrac{1}{4} = \dfrac{3}{4}$

2) $\dfrac{1}{y} + \dfrac{1}{3} = \dfrac{1}{2}$

3) $\dfrac{x}{2x + 4} - \dfrac{1}{x + 2} = 1$

4) $\dfrac{x + 4}{2x - 6} = \dfrac{x}{x - 3} + 2$

5) $\dfrac{2a + 1}{3} - \dfrac{3}{2a - 1} = -2$

6) $\dfrac{a - 2}{a^2 - a - 6} = \dfrac{1}{a^2 - 4} + \dfrac{3}{2a + 4}$

If you had trouble with #6, here is a look-alike:

$$\frac{a}{a + 1} - \frac{a + 1}{a - 4} = \frac{5}{a^2 - 3a - 4}$$

1) Factor: $\qquad \dfrac{a}{a + 1} - \dfrac{a + 1}{a - 4} = \dfrac{5}{(a - 4)\,(a + 1)}$

2) LCD $\quad \left[\dfrac{(a - 4)(a + 1)}{1} \right] \left[\dfrac{a}{a + 1} - \dfrac{a + 1}{a - 4} \right] = \left[\dfrac{(a - 4)(a + 1)}{1} \right] \left[\dfrac{5}{(a - 4)\,(a + 1)} \right]$

3) $\dfrac{(a-4)\cancel{(a+1)}^{1}}{1} \cdot \dfrac{a}{\cancel{\dfrac{a+1}{1}}} - \dfrac{\cancel{(a-4)}(a+1)}{1} \cdot \dfrac{a+1}{\cancel{\dfrac{a-4}{1}}} = \dfrac{\cancel{(a-4)(a+1)}}{1} \cdot \dfrac{5}{\cancel{(a-4)(a+1)}_{1}\ _{1}}$

4) $(a-4)a - [(a+1)(a+1)] = 5$

5) $a^2 - 4a - (\ a^2 + 2a + 1\ ) = 5$

6) $\underline{a^2} - \underline{4a} - \underline{a^2} - \underline{2a} - 1 = 5$

7) $\qquad -6a - 1 = 5$

8) $\qquad -6a - 1 + 1 = 5 + 1$

9) $\qquad -6a - 1 + 1 = 5 + 1$

10) $\qquad -6a = 6$

11) $\qquad -6a - 6 = 0$

12) $\qquad -6(a + 1) = 0$

$-6 \neq 0$	$a + 1 = 0$
	$a + 1 - 1 = 0 - 1$
	$a = -1 \qquad$ Reject

This problem works just like #6, however, it has no solution because if $a = -1$, the denominator of the first fraction on the left would go to zero and cause the fraction to be undefined (meaningless).

Linear Equations and Systems

Let's start by looking at what we call a coordinate plane. The coordinate plane has two axes, a horizontal axis which is called the x-axis and a vertical axis called the y-axis. The axes separate the coordinate plane into four regions called quadrants. The point where the axes intersect, or cross each other, is called the origin. These axes are nothing but number lines and they both share the same zero point or origin (0). Let's look at a drawing of a coordinate plane and how the quadrants are ordered from I to IV.

Graph A

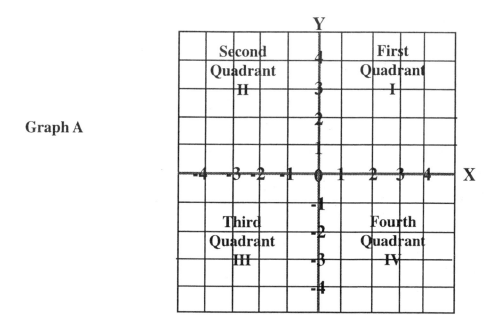

Plotting Points on a Coordinate Plane

Before we go any further, let's stop and discuss what is known as an ordered pair of numbers. When we plot points on a coordinate plane, we must move in a certain order. We always start with the x-axis first. Then we move to the y-axis. This is why we name our coordinate points (**x,** y). To locate and plot the ordered pair (3, 4):

1.   Locate the graph of 3 on the x-axis.

2.   Locate the graph of 4 on the y-axis.

3.   Trace a vertical line through the graph of 3 and a horizontal line through the graph of 4. Make a dot at the point where these lines intersect. This point is called the graph of (3, 4). Remember to always move on the x-axis first, then move on the y-axis second.

In working with graphs, we will sometimes see the terms abscissa and ordinate. These terms refer to the x-axis and the y-axis. The abscissa is referring to the numbers on the x-axis and the ordinate is referring to the numbers on the y-axis.

Now let's look at a drawing of the graph of (3, 4) and a few others.

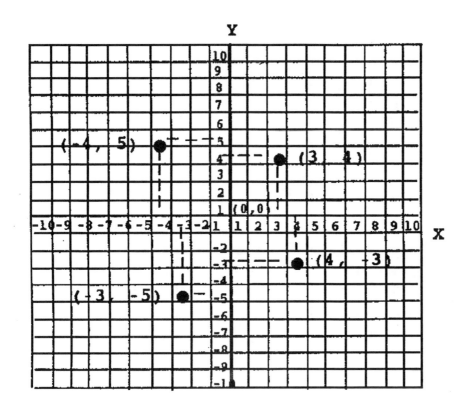

Graph B

Examples of plotting
coordinate points.

1.   (3, 4)
2.   (4, -3)
3.   (-3, -5)
4.   (-4, 5)

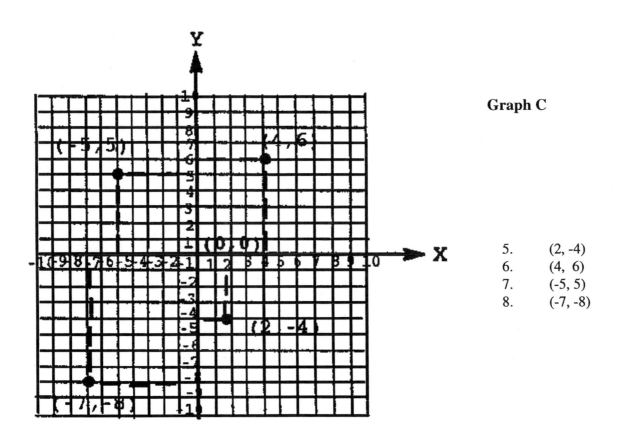

Graph C

5.   (2, -4)
6.   (4, 6)
7.   (-5, 5)
8.   (-7, -8)

Let's Practice #36:   Plot the graph of each ordered pair. Use only one graph for all ten points:

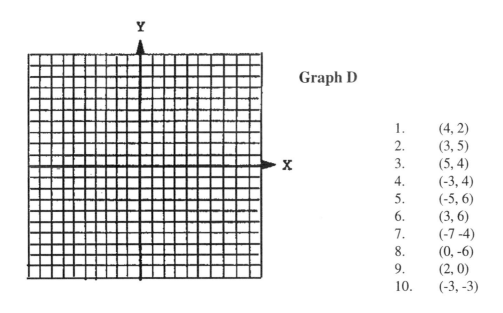

Graph D

1.   (4, 2)
2.   (3, 5)
3.   (5, 4)
4.   (-3, 4)
5.   (-5, 6)
6.   (3, 6)
7.   (-7 -4)
8.   (0, -6)
9.   (2, 0)
10.   (-3, -3)

Equations in Two Variables

Up until now, we have only experienced equations with one variable. Now we will look at some equations having two variables.

Example: $3x + 2y = 10$

Any ordered pair of values of x and y that makes the equation a true statement is considered to be a solution of the equation. For example, if we use the ordered pair (2, 2), the equation above becomes a true statement:

$$3(2) + 2(2) = 6 + 4 = 10$$

If we continue to find more ordered pairs that will produce a solution, we will find what is known as a solution set. This is what we call solving the equation.

Let's solve: $3x + 2y = 10$ using integers for x and y.

1.   First solve the equation for y in terms of x:

$3x + 2y = 10$
$3x - 3x + 2y = 10 - 3x$ Move 3x to the right.

2. $2y = 10 - 3x$

3.　　$\dfrac{2y}{2} = \dfrac{10 - 3x}{2}$　　　　Solve for y

4.　　$y = \dfrac{10}{2} - \dfrac{3x}{2}$

5.　　$y = 5 - \dfrac{3x}{2}$　　or　　$y = 5 - \dfrac{3}{2}x$

6.　　Now let's set up what we call a Table of Values. We do not want to use fractions.

x	$y = 5 - \dfrac{3x}{2}$	Solution (x, y)
1	$5 - \dfrac{3(1)}{2} =$	$\left(1, \dfrac{7}{2}\right)$ No
2	$5 - \dfrac{3(2)}{2} =$	(2, 2) Yes
0	$5 - \dfrac{3(0)}{2} =$	(0, 5)
-1	$5 - \dfrac{3(-1)}{2} =$	$\left(-1, \dfrac{28}{5}\right)$ No
-2	$5 - \dfrac{3(-2)}{2} =$	(-2, 8) Yes

Graph E

Let's graph the equation

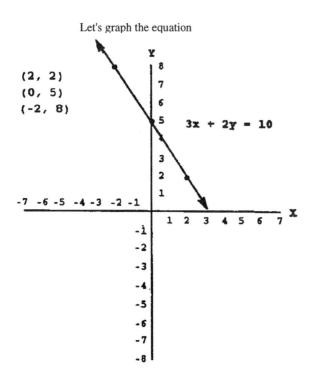

(2, 2)
(0, 5)
(-2, 8)

$3x + 2y = 10$

(Here's how we work)

$5 - \dfrac{3(1)}{2}$

$5 - \dfrac{3}{2}$

$\dfrac{5}{1}\left(\dfrac{2}{2}\right) - \dfrac{3}{2}$

$\dfrac{10}{2} - \dfrac{3}{2} = \dfrac{7}{2}$

$5 - \dfrac{3(2)}{2}$

$5 - \dfrac{6}{2}$

$5 - 3 = 2$

$5 - \dfrac{3(-1)}{5}$

Remember - • - = +

$5 + \dfrac{3}{5}$

$\dfrac{5}{1}\left(\dfrac{5}{5}\right) + \dfrac{3}{5}$

$\dfrac{25}{5} + \dfrac{3}{5} = \dfrac{28}{5}$

$5 - \dfrac{3(-2)}{2}$

$5 + \dfrac{6}{2}$

$5 + 3 = 8$

$5 - \dfrac{3(0)}{2}$

$5 - 0 = 5$

NOTE: See how helpful the identity is for working with fractions. "x" is always independent. We can assign values to x right off the top of our heads, then plug in and work out a value for y.

Before we go any further, I want to mention that we can use fractions when graphing equations, but it is more convenient to plot non-fractional ordered pairs.

Let's look at another example: Graph y = x + 2 in a coordinate plane:

1.   First set up the table of values:

x	y	(x, y)
2	4	(2, 4)
3	5	(3, 5)
0	2	(0, 2)
-2	0	(-2, 0)
-3	-1	(-3, -1)

(You can construct your table to your convenience.)

Graph F

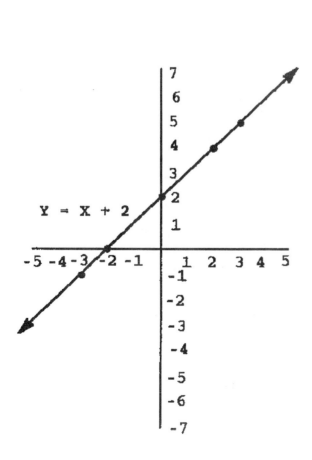

y = x + 2
  = (2) + 2
  = 2 + 2
  = 4

y = x + 2
  = (3) + 2
  = 3 + 2
  = 5

y = x + 2
  = (0) + 2
  = 0 + 2
  = 2

y = x + 2
  = (-2) + 2
  = -2 + 2
  = 0

y = x + 2
  = (-3) + 2
  = -3 + 2
  = -1

As you can see, sometimes it is not necessary to solve an equation for y in terms of **x**. This can be observed in the previous example.

Let's Practice #37:   You will need some grid paper to graph the following problems.

1) $y = 3x - 1$ 2) $x + y = 8$ 3) $x - y = 6$

4) $2y + x = 10$ 5) $4x + 2y = 6$ 6) $6x + 3y = 9$

The Graphic Method

Before we talk about this method, we need to discuss intersecting lines. Intersecting lines are lines that cross or meet at a common point. The place where they cross is called the point of intersection. We have observed that the graph of a linear equation is a straight line. Two straight lines can only intersect at one common point. Now, if we can show that the graphs of two linear equations intersect at a common point on a coordinate plane, we will have found the solution to a system of equations. Finding the solution is called solving the system. A solution of a system of two equations in two variables is an ordered pair of numbers that, when tested, will satisfy both equations. Sometimes we will find that there are some graphs that produce lines that lie in the same plane, but will not intersect. These types are known as parallel lines. Let's look at examples of both types:

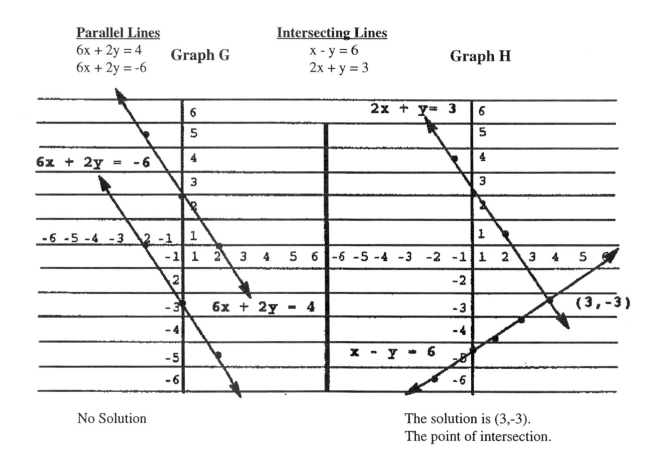

Parallel Lines
$6x + 2y = 4$ **Graph G**
$6x + 2y = -6$

Intersecting Lines
$x - y = 6$ **Graph H**
$2x + y = 3$

No Solution

The solution is (3,-3).
The point of intersection.

94

Let's use the graphic method to solve some systems of equations:

(If we get a fraction for **x** or y, we will throw it out and continue.)

Example #1: Solve: $4x + y = 6$
$x - 2y = 6$

1. Solve each equation for y and set up tables.

$4x + y = 6$	$x - 2y = 6$
$4x - 4x + y = 6 - 4x$	$x - x - 2y = 6 - x$
$y = 6 - 4x$	$-2y = 6 - x$
	$\dfrac{-2y}{-2} = \dfrac{6 - x}{-2}$
	$y = \dfrac{6}{-2} - \dfrac{x}{-2}$
	$y = -3 + \dfrac{x}{2}$

x	y
1	2
2	-2
0	6

x	y
2	-2
0	-3
-2	-4

This is where the two equations intersect. (2, -2)
Use the equation for y each time:

$y = 6 - 4x$ (Let x = 1)

$= 6 - 4(1)$

$= 6 - 4$

$= 2$

$y = -3 + \dfrac{x}{2}$ (Let x = 2)

$= -3 + \dfrac{2}{2}$

$= -3 + 1$

$= -2$

$y = 6 - 4(2)$ (Let x = 2)

$= 6 - 8$

$= -2$

$y = 6 - 4(0)$ (Let x =0)

$= 6 - 0$

$= 6$

$y = -3 + \dfrac{x}{2}$ (Let **x** = 0)

$= -3 - \dfrac{0}{2}$

$= -3$

$y = -3 + \dfrac{x}{2}$ (Let x = -2)

$= -3 + \dfrac{-2}{2}$

$= -3 - 1$

$= -4$

Continue in this method. (About three points is good enough.)
Let's graph the intersection or solution of $4x + y = 6$

$$x - 2y = 6$$

Graph I

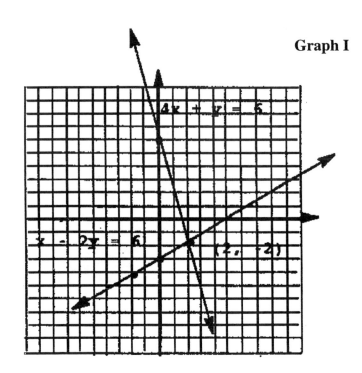

The solution is (2, -2)
NOTE: Since a line has an indefinite number of points, you can give x any value you wish and still get the same results. However, you must plug in and work out the equation each time to get the correct ordered pair.

Example #2:   Solve $y = 2x$
                                   $x + y = 9$

Graph J

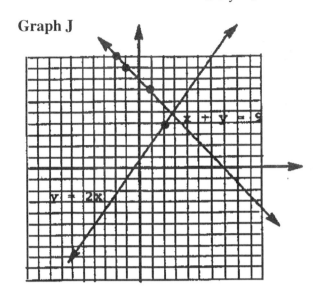

$y = 2x$ $x + y = 9$
                                  $x - x + y = 9 - x$
                                     $y = 9 - x$

x	y		x	y
1	2		1	8
2	4		2	7
0	0		0	9
-1	-2		-1	10
3	6		3	6

Example #3: $x = -3$

Graph K

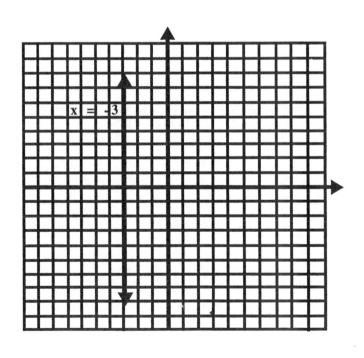

If we would graph both Examples 3 and 4, they would intersect at the point (-3, 3). See below:

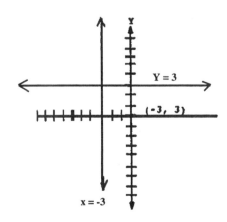

97

Example #4: Graph : y = 3

Graph L

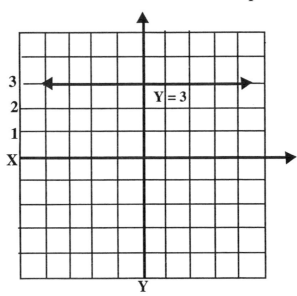

Let's Practice #38:   Solve each system by the graphic method:

1)   x + y = 4   2)   x + 3y = 3
     x - y = 2   2x + 6y = 12

3)   x - y = 5   4)   x + 2y = 2
     x + y = 5   x + y = 1

5)   2x + y = 10   6)   y = 7 - x
     3x - 2y + 8   x - y = 3

If you had trouble solving for y in #1, #3, and #6, here is a look-alike that may be helpful for all three:

Look-alike #1, #3, and #6   x + y = 6

                                       x - y = 8

   x + y = 6   x -y = 8
   x - x + y = 6 - x   x - x - y = 8 - x
   ┌─────────────┐   -y = 8 - x
   │ y = 6 - x │   (-1) (-y) = (8 - x) (-1)
   └─────────────┘
   or y = -x + 6   ┌─────────────┐
                                  │ y = -8 + x │
                                  └─────────────┘
                                  or y = x - 8

Remember we are solving for y **not** -y. Even if you don't find the solution while working out your table, once you draw the graph, the solution will appear at the point of intersection.

98

THE SUBSTITUTION METHOD

To solve a system of linear equations in two variables by using the substitution method:

1. First solve one equation for one of the variables

2. Then substitute the resulting expression in the other equation.

3. Next, solve the resulting equation.

4. Finally, find the corresponding value of the first variable.

Example #1:   Solve the system:
$$x + y = 2$$
$$5x + y = 6$$

NOTE: This method works better when the coefficient of the variable is 1 or -1. Take advantage of this when possible.

1. Solve for y in terms of x in the first equation.
$$x + y = 2$$
$$x - x + y = 2 - x$$
$$y = 2 - x$$

2. Substitute this expression for y in the second equation
$$5x + y = 6$$
$$5x + (2-x) = 6$$

3. Solve for x
$$5x + 2 - x = 6$$
$$4x + 2 = 6$$
$$4x + 2 - 2 = 6 - 2$$
$$4x = 4$$
$$\frac{4x}{4} = \frac{4}{4}$$
$$x = 1$$

4. Find the corresponding value of y by substituting in the expression of Step 1. We could really use either equation to find the corresponding value of x. However, it is more convenient to use $y = 2 - x$, since we know the value of x.
$$y = 2 - x$$
$$y = 2 - 1$$
$$y = 1$$

Let's <u>Check.</u> $x + y = 2$ $5x + y = 6$
                       $1 + 1 = 2$ $5(1) + (1) = 6$
                            $2 = 2$ $5 + 1 = 6$
                                            $6 = 6$

The solution (1, 1) is the answer. These systems are called equivalent systems because they have the same solution set.

NOTE: Be sure to read This example over as many times as needed. Sometimes our results may be fractional ordered pairs.

Let's Practice #39:   Solve each system by the substitution method.

1) $y = 3$ 2) $4x - 2y = -10$
       $x = 15 - 3y$ $x = -3$

3) $x + y = 4$ 4) $4x + 6y = 8$
       $4x + 6 = 22$ $x - 2y = -5$

5) $4x + 6y = 0$ 6) $x - y = 1$
       $x - 6y = -5$ $8x - 2y = 12$

7) $6a - 2b = 28$ 8) $3a - b = 0$
       $a = 2b - 2$ $4a + 6b = 0$

9) $\dfrac{x}{4} - y = 2$
       $x + y = -7$

Remember to treat a and b just as x and y (a, b) as (x, y). Always keep them in the same order as they come in the alphabet.

If you had trouble with #8 and #9, here are some look-alikes:

Look-A-Like #8: $2a - b = 0$
                       $3a + 5b = 0$

1. $2a - b = 0$ 5. $b = 2(0)$ Substitution for a to find b.
       $2a - b + b = 0 + b$
       $2a = b$ **OR** $b = 2a$ 6. $b = 0$

2. $3a + 5 (2a) = 0$ Substitution

3. $3a + 10a = 0$

4. $13a = 0$
       $a = 0$

Let's <u>Check</u>:

$$2a - b = 0 \qquad\qquad 4a + 6b = 0$$
$$2(0) - 0 = 0 \qquad\qquad 4(0) + 6(0) = 0$$
$$0 - 0 = 0 \qquad\qquad\quad 0 + 0 = 0$$
$$0 = 0 \qquad\qquad\qquad 0 = 0$$

The solution is $(0, 0)$

Look-A-Like #9: $\qquad \dfrac{x}{2} - y = 3 \qquad\qquad x + y = -6$

1. $\qquad x + y = -6$
$\qquad\quad x + y - y = -6 - y$
$\qquad\qquad\quad\; x = -6 - y$

2. $\qquad \dfrac{-6 - y}{2} - y = 3 \qquad$ Substitution

3. $\qquad \dfrac{2}{1}\left(\dfrac{-6 - y}{2} - y\right) = \dfrac{3}{1}\left(\dfrac{2}{1}\right)$

4. $\qquad \dfrac{\overset{1}{\cancel{2}}}{1}\left(\dfrac{-6 - y}{\cancel{2}}\right) - \dfrac{2}{1}\left(y\right) = 6$

5. $\qquad -6 - y - 2y = 6$

6. $\qquad -6 - 3y = 6$

7. $\qquad -6 + 6 - 3y = 6 + 6$

8. $\qquad\qquad -3y = 12$

9. $\qquad\qquad \dfrac{-3y}{-3} = \dfrac{12}{-3}$

10. $\qquad y = -4$

11. $\qquad x = -6 - (-4)$

$\qquad\qquad\qquad\qquad\qquad\qquad$ The solution is $(-2, -4)$

12. $\qquad x = -6 + 4$

13. $\qquad x = -2$

Let's <u>Check</u>:

$\frac{x}{2} - y = 3$

$x + y = -6$
$-2 + -4 = -6$
$-6 = -6$

$\frac{-2}{2} - (-4) = 3$

$-1 + 4 = 3$

$3 = 3$

<u>THE ADDITION OR SUBTRACTION METHOD</u>

This method is useful when a term of one equation, not considering the constant term, is the opposite of a term of the other equation.

Example #1:   Solve: $3x - y = 18$
$x + y = 12$

1. You can see that the terms in the variable y are opposites, so let's add them to find an equation in x:

$$3x - y = 8$$
$$\underline{x + y = 12}$$
$$4x \quad\quad = 20$$

2. $\dfrac{4x}{4} = \dfrac{20}{4}$

$x = 5$

3. Now, substitute back into either of the original equations to find a corresponding value of y:

$x + y = 12$

$5 + y = 12$

$5 - 5 + y = 12 - 5$

$y = 7$

The solution set (5, 7)

$$3x - y = 8 \qquad\qquad x + y = 12$$
$$3(5) - 7 = 8 \qquad\quad 5 + 7 = 12$$
$$15 - 7 = 8 \qquad\qquad 12 = 12$$
$$8 = 8$$

Example #3   Solve: $2a - 4b = -8$
                                $2a - 2b = -4$

1. In this particular case, you can see that the terms in the variable a are similar, therefore it will be to our advantage to subtract at this time:

$$2a - 4b = -8$$
$$\underline{(-)2a \;_{(+)}\, 2b =\, _{(+)}\text{-}4}$$
$$-2b = -4$$
$$b = 2$$

Remember to change the signs of
the subtrahend. (Subtraction rule)
* Use substitution to find a.

Here are some rules to go by when using the Addition or Subtraction Method to solve a system of linear equations in two variables:

1. Eliminate one variable by <u>adding</u> similar terms if two are opposites or by <u>subtracting</u> similar terms if two are the same in sign description.

2. Solve the resulting equation for the other variable.

3. Find the corresponding value of the first variable solved by substituting its value in either of the original equations.

Let's Practice #40:

1) $x + y = 4$
      $x - y = 8$

2) $x - 2y = 8$
      $x + 2y = -8$

3) $2x - y = 3$
      $3x + y = 7$

4) $a - b = 3$
      $a + b = 5$

5) $a - 3b = 0$
      $a + 3b = 12$

6) $3a + b = 8$
      $3a - 2b = 2$

7) $2 = -4x + 2y$
      $2 = 4x - 6y$

8) $2x - 3y = 5$
      $8 = 3x - 3y$

9) $2x + y = -7$
      $3 = y - 3x$

10) $3x - \dfrac{1}{4}\, y = 7$

      $2x - \dfrac{1}{4}\, y = 4$

11) $2(a - b) = 14$
      $a + 2b = -2$

12) $3(x - 2y) = 6$
      $2(x + 3y) = -6$

If you had trouble with #8, #10, and #12, here are some look-alikes:

Look-A-Like #8:

$4x - 6y = 10$
$8 = 3x - 6y$

$4x - 6y = 10$
$\underline{3x - 6y = 8}$ (Rewrite in standard order)
$x \qquad = 2$

$4x - 6y = 10$
$4(2) - 6y = 10$ (Substitution)
$8 - 6y = 10$
$8 - 8 - 6y = 10 \quad - 8$
$\qquad - 6y = 2$

$\dfrac{-6y}{-6} = \dfrac{2}{-6}$

$y = -\dfrac{1}{3}$

Solution set $\quad 2, \quad -\dfrac{1}{3}$

Look-A-Like #10:

$2x - \dfrac{1}{3} y = 8$

$\underline{3x - \dfrac{1}{3} y = 3}$

$2x - \dfrac{1}{3} y = 8$

$_{(-)}3x \; _{(+)}\dfrac{-1}{3}y = {}_{(-)}3$

$\overline{-x \qquad = 5}$

$x = -5$

$2x - \dfrac{1}{3} y = 8$

$2(-5) - \dfrac{1}{3} y = 8$

$-10 - \dfrac{1}{3} y = 8$

$-10 + 10 - \dfrac{1}{3} y = 8 + 10$

$- \dfrac{1}{3} y = 18$

$\left(-\dfrac{3}{1}\right) \left(-\dfrac{1}{3}\right) y = \dfrac{18}{1} \left(\dfrac{-3}{1}\right)$

$y = \dfrac{-54}{1} = -54$

$y = -54$

Solution set $(-5, -54)$

Look-A-Like #12: $2(x - 3y) = 8$
$3(x + 2y) = 2$

$(2x - 6y) = 8$
$\underline{(3x + 6y) = 2}$
$5x \qquad = 10$

$$\frac{5x}{5} = \frac{10}{5}$$

$x = 2$

$3x + 6y = 2$
$3(2) + 6y = 2$
$6 + 6y = 2$
$6 - 6 + 6y = 2 - 6$

$6y = -4$
$$\frac{6y}{6} = \frac{-4}{6}$$
$$y = \frac{-2}{3}$$

Solution set $(2, \dfrac{-2}{3})$

MULTIPLICATION WITH THE ADDITION OR SUBTRACTION METHOD

To solve a system of equations such as in the example below, we sometimes have to multiply one or both equations by a number, either negative or positive, to get terms in x or y to be opposites, or the same, before we can start to eliminate:

Example #!: $2x - 6y = 10$
$5x + 2y = 8$

1. Multiply the second equation by 3, so that the terms in y are the same:

$2x - 6y = 10 \quad \rightarrow \quad 2x - 6y = 10$

$3(5x + 2y) = 3(8) \quad \rightarrow \quad 15x + 6y = 24$

2. Add similar terms: $\qquad \qquad \underline{17x \qquad = 34}$

3. Solve the resulting equation: $\qquad \dfrac{17x}{17} = \dfrac{34}{17}$

$\qquad \qquad \qquad \qquad \qquad x = 2$

---

4. Substitute in either of the original equations to find the corresponding value of the first variable found.

$$2x - 6y = 10$$
$$2(2) - 6y = 10$$
$$4 - 6y = 10$$
$$4 - 4 - 6y = 10 - 4$$
$$-6y = 6$$
$$\frac{-6y}{-6} = \frac{6}{-6}$$
$$y = -1$$

Solution set (2, -1)

Now let's look at an example of a system which needs both equations multiplied by some number in order to be solved. We can eliminate either variable first.

1. $\quad 2x - 5y = 15 \quad \rightarrow \quad 3(2x - 5y) = 3(15)$
    $\quad 3x - 2y = 6 \quad \rightarrow \quad 2(3x - 2y) = 2(6)$

We will eliminate x first:

$$6x - 15y = 45$$
$$_{(-)}6x \, _{(+)}- \, 4y = _{(-)}12$$

2. Subtract: $\quad\quad -11y = 33$

3. Solve for y: $\quad \dfrac{-11y}{-11} = \dfrac{33}{-11}$

$$y = -3$$

4. Substitute in either original equation to find the other variable:
$$6x - 15y = 45$$
$$6x - 15(-3) = 45$$
$$6x + 45 = 45$$
$$6x + 45 - 45 = 45 - 45$$
$$6x = 0$$
$$x = 0$$

Solution set (0, -3)

Let's Practice #41:

1) $\quad 2x + 3y = 0$ 2) $\quad 3x - 8y = 7$
    $\quad x + 5y = 7$ $\quad x + 2y = 7$

3) $\quad 3x + 2y = -12$ 4) $\quad 2a + b = -2$
    $\quad 5x - y = -20$ $\quad 4a + 3b = -6$

5) $3a + 2b = 13$
       $2a + 3b = 12$

6) $2a - 5b = 15$
       $3a - 2b = 6$

7) $3c + 4d = -2$
       $2c - 3d = 10$

8) $2c + 3d = 4$
       $3c - d = 6$

9) $3c - 4d = 1$
       $2c + d = -3$

10) $3x + 2y = -4$
       $2x - 3y = -7$

11) $3a + 7b = -18$
       $5a - 2b = -30$

12) $5a - 2b = 7$
       $2a + 7b = -5$

Slope of Line

When we speak of the slope of a line, let's think of a hill and, as we know, some hills are steeper than others. If we would stand at the foot of a hill and look to its top, we would see that there is a certain steepness. This steepness is called the slope of the hill.

This is something all skiers are familiar with because the slope of a hill will determine the speed in which one will be able to travel. There is a definite ratio between the vertical rise and the horizontal run, similar to the slope of a hill. To describe the steepness, or slope of a line, you may choose any two points on the line, then count the units in the rise and run and then calculate their ratio.

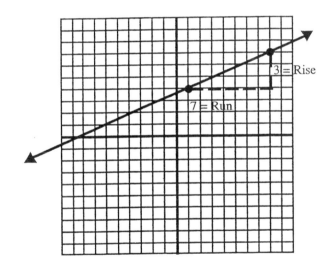

$$\text{Slope} = \frac{\text{Rise}}{\text{Run}} = \frac{3}{7}$$

Example #2

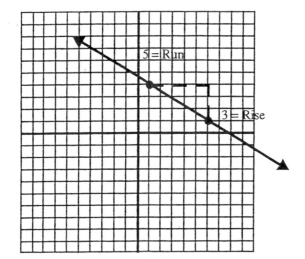

$$\text{Slope} = \frac{\text{Rise}}{\text{Run}} = \frac{-3}{5}$$

We do not have to count the rise and the run to find the slope of a line. The coordinates of a pair of points on a line can be used to calculate the slope of the line.

$$\text{Slope} = \frac{\text{rise}}{\text{run}} = \frac{\text{vertical change}}{\text{horizontal change}} = \frac{\text{change in y}}{\text{change in x}}$$

Let's say that P1 (x_1, y_1) and P2 (x_2, y_2) are different points on a line, the following is an illustration of the slope formula:

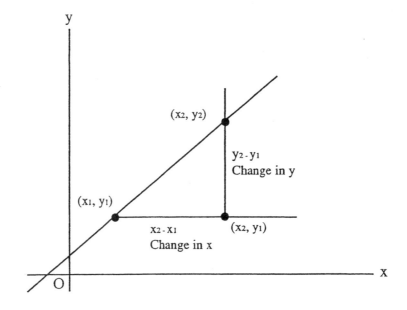

$$\text{Slope} = \frac{\text{Change in y}}{\text{Change in x}} = \frac{y_2 - y_1}{x_2 - x_1}$$

Example #1: Find the slope of the line whose equation is:

$$2x + 3y = 6$$

Let's make a table and assign $x = 0$ and get a corresponding value for y. Then we will do the same with $y = 0$:

x	y
0	2
3	0

$2x + 3y = 6$

$2(0) + 3y = 6$

$0 + 3y = 6$

$3y = 6$

$\dfrac{3y}{3} = \dfrac{6}{3}$

$y = 2$

$2x + 3y = 6$

$2x + 3(0) = 6$

$2x = 6$

$\dfrac{2x}{2} = \dfrac{6}{2}$

$x = 3$

$(x_1, y_1)\ (x_2, y_2)$

Points $(0, 2)$ and $(3, 0)$

Now let's use the slope formula:

Slope $= \dfrac{y_2 - y_1}{x_2 - x_1}$ or slope $= \dfrac{y_1 - y_2}{x_1 - x_2}$

$= \dfrac{0 - 2}{3 - 0}$ $= \dfrac{2 - 0}{0 - 3}$ *They both mean the same

$= \dfrac{-2}{3}$ $= \dfrac{2}{-3}$

It will help a lot to label which points that you want to be (x_1, y_1) and (x_2, y_2). It really doesn't matter as long as you are consistent and make sure that wherever x_1 is there will be a y_1 with it and likewise with x_2 and y_2.

Example #2: Given the coordinates of two points on a line, find the slope of the line:

x_1, y_1 x_2, y_2

$(3, 6)$ $(5, 2)$

Slope $= \dfrac{y_2 - y_1}{x_2 - x_1} = \dfrac{2 - 6}{5 - 3} = -\dfrac{4}{2} = -2$

Now let's keep in mind that a basic property of a line is that its slope is constant. Since a line has many points, we should be able to use any two points on it and still get the same slope as our result.

Let's look at Example #2 again to show how we can label the points that we want to be (x_1, y_1) and (x_2, y_2) in any way we desire.

Find the slope of the line:

$$x_2, \ y_2 \qquad x_1, \ y_1$$
$$(3, \ 6) \ \text{ and } \ (5, \ 2)$$

$$\text{Slope} = \frac{y_1 - y_2}{x_1 - x_2} = \frac{2 - 6}{5 - 3} = -\frac{4}{2} = -2$$

As you can see, we still get the same results.

Let's Practice #42: Find the slope of a line whose coordinates of two points of the line are given:

1) (7, 3), (-2, 3)

2) (4, 6), (4, -3)

3) (3, 8), (5, 4)

4) (3, -3), (6, -5)

5) (5, 4), (7, 3)

6) (4, 2), (2, -5)

Find the slope of each line whose equation is given:

7) $y = 4x + 8$

8) $x - y = 6$

9) $y = 2x - 3$

10) $2x + y = 6$

11) $2x - 3y = 12$

12) $y = 4x + 5$

If you had trouble with #8 and #11, here are some look-alikes:

Look-A-Like #8: $x - y = 8$

Let's look at $x = 0$ and $y = 0$ to get the points we need:

x	y
0	-8
8	0

$$x_1, \ y_1 \qquad x_2, \ y_2$$
$$\text{Points} \quad (0, -8) \ \text{ and } \ (8, 0)$$

$$\text{Slope} = \frac{y_2 - y_1}{x_2 - x_1} = \frac{0 - (-8)}{8 - 0} = \frac{0 + 8}{8} = \frac{8}{8} = 1$$

Remember: $- \bullet - = +$

Look-A-Like#11: $3x - 2y = 6$

Let's look at $x = 0$ and $y = 0$ to get the points we need:

x	y
0	-3
2	0

x_1, y_1 x_2, y_2

$(0, -3)$ $(2, 0)$

$3x - 2y = 6$

$3(0) - 2y = 6$

$0 - 2y = 6$

$$\frac{-2y}{-2} = \frac{6}{-2}$$

$y = -3$

$3x - 2y = 6$

$3x - 2(0) = 6$

$3x - 0 = 6$

$3x = 6$

$$\frac{3x}{3} = \frac{6}{3}$$

$x = 2$

$$\text{Slope} = \frac{y_2 - y_1}{x_2 - x_1} = \frac{0 - (-3)}{2 - 0} = -\frac{3}{2}$$

NOTE: If the slope formula results in an expression like this $\frac{5}{0}$, the slope does not exist.

Remember, we cannot use zero as a divider, so this is a NO SLOPE situation.

THE SLOPE-INTERCEPT FORM OF A LINEAR EQUATION

Before we start, let's look at a couple of rules:

1.   For every real number m, the graph in a coordinate plane of the equation:

$$y = mx$$

is the line that has slope m and passes through the origin.

An example of this rule is: $y = 3x$

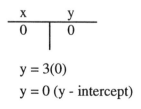

x	y
0	0

$y = 3(0)$

$y = 0$ (y - intercept)

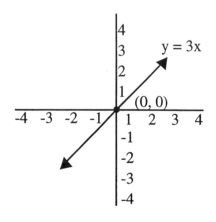

Remember the y-coordinate of a point where a graph intersects the y-axis is called the y-intercept of the graph.

Let's look at another equation: y = 3x + 5 (replace x with 0 in the equation:)   y = 3(0) + 5

$$y = 0 + 5$$
$$y = 5$$

This is the y-intercept of the graph.

The y-intercept is simply the point on the y-axis where the graph touches, or cuts, through.

NOTE: For all real numbers m and b, the graph in a coordinate plane of the equation:

$$y = mx + b$$

is the line whose slope is m and whose y-intercept is b. This is called the slope-intercept form of an equation of a line.

Example #1:   Write a linear equation in standard form whose graph has the given slope and y-intercept:

(a) m = 3, b = 5

Write the slope-intercept form of an equation:

$$y = mx + b$$

Then substitute for m and b:

$$y = mx + b$$

$$y = 3x + 5$$

Now transform y = 3x + 5 into an equivalent equation in the form of: ax + by = c, where a, b, and c are integers.

y = 3x + 5

y - 3x = 3x - 3x + 5   (Move 3x to the left

y - 3x = 5   side) (Rearrange the terms)

$$\boxed{-3x + y = 5}$$ (Answer)

Example #2:   Use only the y-intercept and slope to graph the equation 2x + 3y = 6.

Let's solve for y by transforming the given equation into the form: y = mx + b

$$2x + 3y = 6$$

1. Move 2x to the right side:

$$2x - 2x + 3y - 6 - 2x$$
$$3y = 6 - 2x$$

2. Divide both sides by 3:

$$\frac{3y}{3} = \frac{6 - 2x}{3}$$

$$y = \frac{6}{3} - \frac{2x}{3}$$

3. Rearrange terms:

$$y = 2 - \frac{2}{3}x$$

$$y = -\frac{2}{3}x + 2$$

Since the y-intercept is 2, plot (0, 2). Since the slope is $-\frac{2}{3}$ measure 3 units to the right of (0, 2) and 2 units down to locate a second point. Draw the line through the two points:

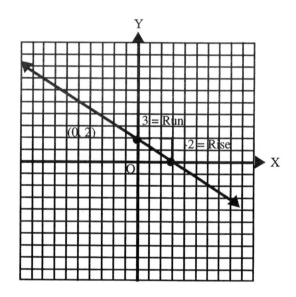

Remember, any time we go down on the y-axis, this is considered to be in the negative direction no matter where we are starting from. This is why the rise is -2 in the figure above. The same goes when we go left on the x-axis.

Let's Practice #43:

Write a linear equation in standard form whose graph has the given slope and y-intercept. Remember to use the equation y = mx + b and substitute for m and b. Then work out to get the form ax + by = c.

1) $m = 5, \ b = -\dfrac{2}{3}$ 2) $m = -\dfrac{2}{5}, \ b = -4$

3) $m = 0, \ b = -3$

Use only the y-intercept and slope to graph the following equations: (Remember to solve for y to get the equation in the form y = mx + b.)

      4)   2x + y = -4   5)   2y - 3x = 6

      6)   -x + 2y = 6

    If you had any trouble with #3 and #5, here are some look-alikes:

      Look-alike #3:   m = 0, b = -5

  1.   y = mx + b
       y = 0x + -5   (Substitute for m and b)
       y = -5   **NOTE:** 0x = 0

Look-A-Like #5:   3y - 2x = 6

1.   Solve for y,
     move 2x to the right side: 3y - 2x + 2x = 6 + 2x
                               3y = 2x + 6

2.   Solve for y: $\dfrac{3y}{3} = \dfrac{2x + 6}{3}$

                  $y = \dfrac{2}{3}x + 2$

The slope is $\dfrac{2}{3}$ and the y-intercept is 2. Since the y-intercept is 2, plot (0, 2). Since the slope is $\dfrac{2}{3}$, measure 3 units to the right of (0, 2) and 2 units up to locate a second point. Draw the line through the two points.

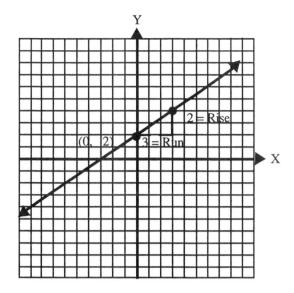

Determining the Equation of a Line

We have just shown how to write an equation of a line by using the slope and the y-intercept. Now we are going to show how to write an equation of a line by using the slope and one point on the line or two points on the line.

Example #1: The line shown below has slope $\frac{3}{4}$ and passes through the point (-3, -1). Since the slope-intercept form of an equation is y = mx + b, the slope-intercept form of this equation is:

$$y = \frac{3}{4} x + b,$$

Since the point (-3, -1) is on the line, its coordinates satisfy the equation. We can substitute to find the value of b:

$$x = -3 \text{ and } y = -1$$

$$y = \frac{3}{4} x + b$$

$$-1 = \frac{3}{4} \left(-\frac{3}{1} \right) + b \qquad \textbf{NOTE:} \quad -3 = -\frac{3}{1}$$

$$-1 = \frac{-9}{4} + b \qquad \text{Solve for b.}$$

$$-1 + \frac{9}{4} = \frac{-9}{4} + \frac{9}{4} + b$$

$$-\frac{4}{4} + \frac{9}{4} = b \qquad \textbf{NOTE:} \qquad -\frac{4}{4} \quad \text{is the same}$$
$$\phantom{-\frac{4}{4} + \frac{9}{4} = b \qquad \textbf{NOTE:} \qquad -\frac{4}{4} \quad} \text{as -1}$$

$$\text{y- intercept} \qquad \qquad \frac{5}{4} = b$$

Remember the identity element.

Since we have found the y-intercept, we can write an equation of the line with slope $\frac{3}{4}$ and passing through (-3, -1).

Answer: $y = \frac{3}{4}x + \frac{5}{4}$ See graph below

Write this answer in standard form $-\frac{3}{4}x + y = \frac{5}{4}$

$\left(\frac{4}{1}\right)$ $\left(\frac{-3}{4}x + y = \frac{5}{4}\right)$ $\left(\text{Multiply by } \frac{4}{1}\right)$

   $-3x + 4y = 5$ OR $3x - 4y = -5$ (Standard form)

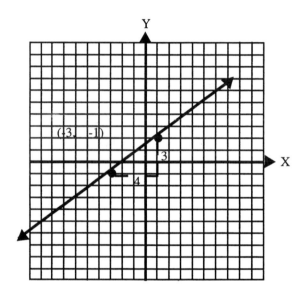

NOTE: We multiply for $\frac{4}{1}$ to clear the denominators.

NOTE:
Always read through the examples several times very carefully, then try to work them for yourself. This will help your understanding.

Example #2:   Now, let's look at how we write an equation of a line passing through <u>two</u> <u>points</u>.

In order to do this, we must recall the formula for finding the slope:

$$\text{Slope} = \frac{y_2 - y_1}{x_2 - x_1}$$

$$\begin{array}{cc} x_1, y_1 & x_2, y_2 \end{array}$$
Since the points are (-2, 1) and (3, 6), we substitute in the formula:

$$\text{Slope} = \frac{6 - 1}{3 - (-2)} = \frac{5}{3 + 2} = \frac{5}{5} = 1$$

Since, in the formula y = mx + b, the m represents the slope, we replace m with 1. The slope-intercept form of the equation is:

$$y = 1x + b \quad \text{OR} \quad y = x + b \text{ (Standard form)}$$

Now let's choose one of the given points, either (-2, 1) or (3, 6) to substitute into the equation and find the value of b:

Let's start with (-2, 1)   (3, 6)

     y = x + b   y = x + b

     1 = -2 + b   6 = 3 + b

     1 + 2 = -2 + 2 + b   6 - 3 = 3 - 3 + b

     3 = b   3 = b

You see, either point will work.

Remember: $\therefore$ means therefore.

$\therefore$ An equation of the line is:

     y = 1x + 3

     $\boxed{y = 1x + 3}$ **(Answer)**

Write this equation in standard form:

     y = x + 3

     y - x = x - x + 3   (Move x to the left side)

     -x + y = 3   (Rearrange terms to look standard.)

     -1(-x + y) = 3   (Multiply the whole equation by -1)

     x - y = -3   (Answer)

NOTE: Either $-x + y = 3$ OR $x - y = -3$ is a correct answer. It is more standard to have the ax term positive. Remember the equation $ax + by - c$, where a, b, and c are integers. When the first term is negative, always multiply the entire equation by -1.

Let's practice #44:

Write in standard form an equation of the line that has the given slope and passes through the given point:

1) $m = 3$; (2, 3)   2) $m = -\dfrac{1}{2}$; (3, -2)

3) $m = 0$; (4, -2)   4) $m = \dfrac{5}{4}$; (1, -5)

Write, in standard form, an equation of the line passing through the given points:

5) (-4, -2) (0, -5)   6) (2, -1) (-3, 4)

7) (6, 0) (2, 1)   8) (2, -2) (5, 6)

Solving Inequalities

When we solve an inequality, we usually try to transform it into a simple equivalent inequality whose solution, or solutions, can be seen at a quick glance. We are going to assume that the domain (replacement values) of all variables is the set of real numbers unless otherwise mentioned. In short, we will solve inequalities very similarly to the way we solve our normal equations.

Before we start, let's review our symbols used in working with inequalities.

1.	$>$	Greater than
2.	$<$	Less than
3.	$\geq$	Greater than or equal to
4.	$\leq$	Less than or equal to

Example #1:

$3x - 2 < 8 + x$ Solve and graph

$3x - x - 2 < 8 + x - x$ Move x to the left

$2x - 2 < 8$

$2x - 2 + 2 < 8 + 2$

$2x < 10$

$\dfrac{2x}{2} < \dfrac{10}{2}$

$x < 5$

118

Start very close to 5, but not on 5.   (That's what the open bubble means)

Example #2: $2(x - 3) > 3(x + 2) - 3$

$2x - 6 > 3x + 6 - 3$ (Move the parentheses)

$2x - 6 > 3x + 3$

$2x - 3x - 6 > 3x - 3x + 3$

$-x - 6 > 3$

$-x - 6 + 6 > 3 + 6$

$-x > 9$

$(-1)(-x) < 9(-1)$

$x < -9$

When we multiply or divide both sides by -1, we reverse the order of the inequality. Don't forget this as you, work.

NOTE: The larger a negative number, the less its value.

Let Practice #45:   Solve each inequality:

1) $3x < 2x + 8$ 2) $4x - 3 > 3x + 6$

3) $2y - 3 < 4y + 5$ 4) $3(x + 5) \leq 6$

5) $8 < 4(2 - x)$ 6) $2(x - 6) < 3(x - 2)$

Solve each inequality and draw its graph:

7) $3(x - 1) \geq \dfrac{3}{4} x$ 8) $\dfrac{2}{3} x \leq 4(x - 5)$

9) $2(x - 2) - 4 \leq 3(x + 1) + 3$

10) $3(2y - 1) - 2(y + 1) > 3y + 8$

NOTE: Don't forget to combine like terms in #10.

If you had trouble with #6 and #8, here are some look-alikes:

Look-alike #6 $3(x - 2) < 2(x - 4)$

$3x - 6 < 2x - 8$ (Move the parentheses)

$3x - 6 + 6 < 2x - 8 + 6$ (Move 6 to the right side)

$3x < 2x - 2$

$3x - 2x < 2x - 2x - 2$

$x < -2$

Look-alike #8 $\dfrac{3}{4} x \leq 4(x - 2)$

$\dfrac{3}{4} x \leq 4x - 8$

$\dfrac{3}{4} x - 4x \leq 4x - 4x - 8$

$\dfrac{3}{4} x - \dfrac{4x}{1} \left(\dfrac{4}{4}\right) \leq -8$ (Get common denominator and combine terms.)

$\dfrac{3}{4} x - \dfrac{16x}{4} \leq -8$

$-\dfrac{13x}{4} \leq -8$

$\left(-\dfrac{4}{13}\right) \left(-\dfrac{13x}{4}\right) \geq \dfrac{8}{1} \left(-\dfrac{4}{13}\right)$ (Reverse the inequality)

$x \geq \dfrac{32}{13}$ **OR** $2\dfrac{6}{13}$

This means starting at 2 and go to the right:

-7 -6 -5 -4 -3 -2 -1 0 1 2 3 4 5 6 7

NOTE: When the bubble is shaded in, it means to include the number where we start. If the bubble is not shaded, it means to get very close to the number where we start, but do not include it.

Our next topic will be the first step to leading us into what most students dread. This topic is going to lead us into word problems.

TRANSLATING WORD PHRASES INTO MATHEMATICAL EXPRESSIONS OR EQUATIONS

Under this topic, we will represent each word phrase by a variable expression.

Example #1;   Four more than twice a number.

Solution: Since we do not know the number, we will let x represent the number. Now, to show four more than twice the number we need 2x + 4.

Before we go any further, let's make a few important notes that we must keep in mind:

1.   Of   -   means time

2.   Is   -   means equal to

3.   More   -   means + (plus)

4.   Less   -   means - (minus)

5.   Sum   -   means + (add)

6.   Difference   -   means - (subtract)

7.   Product   -   means x (multiply)

8.   Quotient   -   means   ÷ (divide)

9.   Increased   -   means   + (Add)

10.   Decreased   -   means - (subtract)
      or diminished

11.   Exceed   -   means to add on (plus)

Now, let's look at **Example #2:**

Represent each sentence by an equation. Use n to represent the number that we are referring to. It really doesn't matter which variable we assign the number to be, but to be consistent, use the same variable throughout this exercise, whichever your choice may be:

Example #2:   Six less than the number is 12.

Solution: $n - 6 = 12$

Let's Practice #46:

Represent each word phrase by a variable expression. Use n for the variable:

1)   Three times a number.

2)   The number increased by seven.

3)   The product of five and a number.

4)   The number is decreased by 2.

5)   Eight less than half the number.

6)   Four greater than the number

Represent each sentence by an equation. Use n for the variable:

7)   Six more than the number is 50.

B)   Nine less than the number is 72.

9)   The sum of seven and three times the number is twenty-five.

10)   Five less than the number is 12.

11)   Eight less than twice the number is 20.

12)   If one is subtracted from five times the number, the difference is fourteen.

Let's Practice #47:

Represent each sentence by an equation and solve.

1) The sum of twice a number and 84 is 104.   Find the number

2) Six times a number, increased by 7 is 55   Find the number

3) Twice a number, decreased by 35, is -5.   Find the number.

4) Three times a number, decreased by 6, is 48.   Find the number.

5) Four times a number, decreased by 46, is -10.   Find the number.

6) The sum of 17 and three times a number is 80.   Find the number.

INTEGER PROBLEMS

Before we start, let's talk about the different types of integers that we will be working with. There are three that we work with most of the time they are consecutive; consecutive evens; and consecutive odds.

Here is how we write consecutive integers:

Let n be the 1st. (first)

Let n + 1 be the 2nd (second)

Let n + 2 be the 3rd (Third)   **NOTE:** $n + 1 + 1 = n + 2$

Let n + 3 be the 4th (Fourth) $n + 2 + 1 = n + 3$

As you can see, when we say consecutive, we mean one after the other. This is why we count by ones.

Now let's look at the even integers and how we write them:

Let n be the 1st (first)

Let n + 2 be the 2nd (second)   (Remember to count by two's on consecutive, even numbers.)

Let n + 4 be the 3rd (third)

Now, when we look at consecutive odd integers, we count the same way as we do for consecutive even integers. The reason for this is that all even integers are exactly two units apart. The same goes for odd integers. Take a look at the number line below.

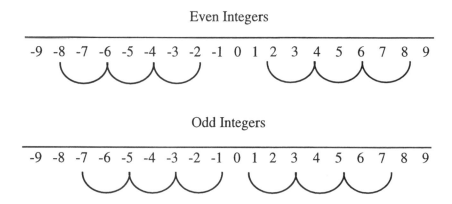

So, as you can see, if we start with an even integer, the next consecutive will always be even, if we add two each time. The same situation applies to the odd integers.

Let's look at how we write the consecutive odd integers:

Let n be the 1st (First)

Let n + 2 be the 2nd (Second)   Remember to count by two's on consecutive odd integers.

Let n + 4 be the 3rd (Third)

Now, let's work some examples using consecutive integers.

Example #1: Find two consecutive integers whose sum is 27.

Let n be the 1st.

Let n + 1 be the 2nd

Relationship: Their sum is 27.

$$n + (n + 1) = 27$$

$$n + n + 1 = 27$$

$$2n + 1 = 27$$

$$2n + 1 - 1 = 27 - 1$$

$$2n = 26$$

$$\frac{2n}{2} = \frac{26}{2}$$

$$n = 13 \ (1st)$$

$$n + 1 = 13 + 1 = 14 \ (2nd)$$

13, 14 **(Answers)**

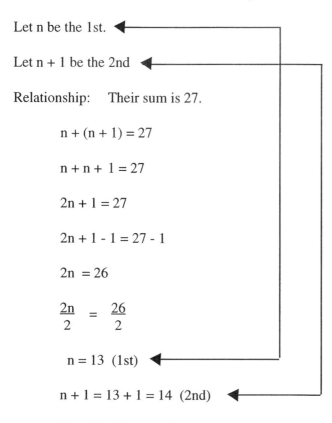

Remember, if the word consecutive is used and there is no mention of odd or even, we count by one's: x, x + 1, x + 2, and so on.

Example #2:   Find three consecutive **even** integers whose sum is 48.

NOTE: It may seem hard, but always write the Let statements.

Let n be the 1st consecutive even integer.

Let n + 2 be the 2nd consecutive even integer.

Let n + 4 be the 3rd consecutive even integer.

Relationship: Their sum is 48.

$$n + (n + 2) + (n + 4) = 48$$

$$n + n + 2 + n + 4 = 48$$

$$3n + 6 = 48$$

$$3n + 6 - 6 = 48 - 6$$

$$3n = 42$$

$$\frac{3n}{3} = \frac{42}{3}$$

$$n = 14$$

Now, by looking back at the Let statements, we can complete the problem:

Since n = 14 (1st Integer)

n + 2 = 14 + 2 = 16 (2nd Integer)

n + 4 = 14 + 4 = 18 (3rd Integer)

14, 16, 18 is the answer.

NOTE: The answer is incomplete, If we don't find all three integers. We can not stop at the first value of n.

Example #3:   Find four consecutive <u>odd</u> integers whose sum is 96.

Let n be the 1st consecutive odd integer.

Let n + 2 be the 2nd consecutive odd integer.

Let n + 4 be the 3rd consecutive odd integer.

Let n + 6 be the 4th consecutive odd integer.

Relationship: Their sum is 96.

$$n + (n + 2) + (n + 4) + (n + 6) = 96$$

$$n + N + 2 + n + 4 + n + 6 = 96$$

$$4n + 12 = 96$$

$$4n + 12 - 12 = 96 - 12$$

$$4n = 84$$

$$\frac{4n}{4} = \frac{84}{4}$$

$$n = 21$$

Now, by looking back at the Let statements, we can complete the problem: This is a <u>must.</u>

$$n = 21$$

$$n + 2 = 21 + 2 = 23$$

$$n + 4 = 21 + 4 = 25$$

$$n + 6 = 21 + 6 = 27$$

21, 23, 25 27 is the answer.

NOTE: Never jump to conclusions, always read the word problem at least twice. Sometimes after finding the first answer, it will help you complete the main answer.

Now, let's look at consecutive integer problems with another dimension:

Example #4:

The sum of the least and greatest of four integers is 11. What are the integers? (Remember to count by one's.)

(LEAST)   Let n be the 1st integer

Let n + 1 be the 2nd integer

Let n + 2 be the 3rd integer

(GREATEST)   Let n + 3 be the 4th integer

Relationship: The sum of the least and the greatest is 11.

$$n + (n + 3) = 11$$

$$n + n + 3 = 11$$

$$2n + 3 = 11$$

$$2n + 3 - 3 = 11 - 3$$

$$2n = 8$$

$$\frac{2n}{2} = \frac{8}{2}$$

$$n = 4$$

Looking back at the Let statements:

$$n = 4$$

$$n + 1 = 4 + 1 = 5$$

$$n + 2 = 4 + 2 = 6$$

$$n + 3 = 4 + 3 = 7$$

4, 5, 6, 7, is the answer.

Example #5:

The sum of the least and greatest of three consecutive <u>odd</u> integers is 34. Find the middle integer.

(LEAST)   Let n be the 1st consecutive odd integer.

Let n + 2 be the 2nd consecutive odd integer.

Let n + 4 be the 3rd consecutive odd integer.

Relationship: Greatest + Least = 34

$$(n + 4) + n = 34$$

$$n + 4 + n = 34$$

$$2n + 4 = 34$$

$$2n + 4 - 4 = 34 - 4$$

$$2n = 30$$

$$\frac{2n}{2} = \frac{30}{2}$$

1st integer: n = 15

2nd integer: n + 2 = 15 + 2 = 17

3rd integer: n + 4 = 15 + 4 = 19

(Middle integer) 17 (answer)

Let's Practice #48:

1)   Find two consecutive integers whose sum is 35.

2)   Find 3 consecutive <u>odd</u> integers whose sum is 51.

3)   Find four consecutive <u>even</u> integers whose sum is 28.

4)   Find three consecutive integers such that the sum of the greatest and twice the least is 32.

*5)   Find four consecutive integers such that the sum of the two greatest subtracted from twice the sum of the two least is 41.

*6)   Find four consecutive integers such that four times the third decreased by twice the fourth is 12.

*7)   Find four consecutive odd integers such that the third is the sum of the fourth and twice the second.

NOTE: When there is a figure involved, always draw the figure first and label it.

*8)   The measure in meters of two adjacent sides of a rectangle are consecutive odd integers. The perimeter is 96m. What are the dimensions? (What is its length and width?)

You may need a little help setting up this problem:

n + 2 (length)

```
 ┌─────────────────┐
 │                 │
n│ (width)         │
 │                 │
 │                 │
 └─────────────────┘
```

Perimeter: The distance around a figure.

The formula for perimeter is: P - 2L + 2W

Let's turn the formula around:

$$2L + 2W = P$$

$$2(n + 2) + 2(n) = 96 \text{ (Substitution)}$$

The rest is left up to you. Get to it, because you can do it.

If you had trouble with #5, #6, and #7, here are some look-alikes:

Look-A-Like #5:

Find four consecutive integers such that the sum of the two greatest subtracted from twice the sum of the two least is 5.

(LEAST)   Let n be the 1st consecutive integer.

(LEAST)   Let n + 1 be the 2nd consecutive integer.

(GREATEST)   Let n + 2 be the 3rd consecutive integer.

(GREATEST)   Let n + 3 be the 4th consecutive integer.

Relationship: The sum of the two greatest subtracted from twice the sum of the two least is 5.

TWICE TWO LEAST   —   **SUM OF TWO GREATEST**

$2[n + (n + 1)]$ — $[(n + 2) + (n + 3)] = 5$

(Remember how to use ( ) and [ ].)

$2[n + n + 1]$ — $[n + 2 + n + 3] = 5$

(Work inside first)

$2(2n + 1)$ — $(2n + 5) = 5$

(You do not need [ ] here.)

$$4n + 2 - 2n - 5 = 5$$

$$2n - 3 = 5$$

$$2n - 3 + 3 = 5 + 3$$

$$2n = 8$$

$$\frac{2n}{2} = \frac{8}{2}$$

$$n = 4$$

Remember, always give a complete answer.

Looking back at the Let statements: $n = 4$

$$n + 1 = 4 + 1 = 5$$

$$n + 2 = 4 + 2 = 6$$

$$n + 3 = 4 + 3 = 7$$

4, 5, 6, and 7 is the answer.

Look-A-Like #6:

Find four consecutive integers such that four times the third decreased by twice the fourth is 4.

Let's use x for our variable this time:

Let x be the 1st consecutive integer.

Let x + 1 be the 2nd consecutive integer.

Let x + 2 be the 3rd consecutive integer.

Let x + 3 be the 4th consecutive integer.

Relationship:

$$4(\text{Third}) - 2(\text{Fourth}) = 4$$

$$4(x + 2) - 2(x + 3) = 4$$

$$4x + B - 2x - 6 = 4$$

$$2x + 2 = 4$$

$$2x + 2 - 2 = 4 - 2$$

$$2x = 2$$

$$\frac{2x}{2} = \frac{2}{2}$$

$$x = 1$$

1st integer:	x = 1
2nd integer:	x + 1 = 1 + 1 = 2
3rd integer:	x + 2 = 1 + 2 = 3
4th integer:	x + 3 = 1 + 3 - 4

1, 2, 3, 4 is the answer.

Look-A-Like #7:

    Find four consecutive odd integers such that the fourth is the sum of the third and twice the second.

Let a be the 1st consecutive odd integer.

Let a + 2 be the 2nd consecutive odd integer.

Let a + 4 be the 3rd consecutive odd integer.

Let a + 6 be the 4th consecutive odd integer.

Relationship: The fourth is the sum of the third and twice the second.

$$a + 6 = (a + 4) + 2(a + 2)$$

$$a + 6 = a + 4 + 2a + 4$$

$$a + 6 = 3a + 8$$

$$a + 6 - 6 = 3a + 8 - 6 \qquad \text{(Remember the}$$
$$\text{(+) + (-) rule)}$$

$$a = 3a + 2$$

$$a - 3a = 3a - 3a + 2$$

$$-2a = 2$$

$$\frac{-2a}{-2} = \frac{2}{-2}$$

$$a = -1$$

1st integer:	$a = -1$
2nd integer:	$a + 2 = -1 + 2 = 1$
3rd integer:	$a + 4 = -1 + 4 = 3$
4th integer:	$a + 6 = -1 + 6 = 5$

    -1, 1, 3, 5 is the answer.

AGE PROBLEMS

In order to do age problems, we must always be concerned with present, past and future. We must remember that <u>years from now or later,</u> means that we must add on and <u>years ago</u> means that we must subtract.

Example #1: John is three years older than Bill. In three years their age will total 25. How old is each now?

NOTE: Be sure to read all of the information carefully, at least twice.

Let's think before we start. We do not know how old Bill is, but we do know that John is three years older than whatever age Bill is. Now, which one do we know the least about? The answer is Bill. Since we know the least about Bill's age, let's assign the first variable to be equal to Bill's age.

(Present)   Let x be Bill's age.

Let $x + 3$ be John's age.

(Future)   Three years later:

$x + 3$ - Bill's age.

$(x + 3) + 3 = x + 3 + 3 = x + 6 = $ John's age

At this time they totaled 25.

Relationship: Bill's age + John's age = 25

$(x + 3) + (x + 6) = 25$ (Substitution)

$x + 3 + x + 6 = 25$

$2x + 9 = 25$

$2x + 9 - 9 = 25 - 9$

$2x = 16$

$\dfrac{2x}{2} = \dfrac{16}{2}$

$x = 8$ (Bill's age)

$x + 3 = 8 + 3 = 11$ (John's age)

Example #2:

Mrs. Brown is 3 times as old as her son. Ten years ago she was 5 times as old as her son was then. Find each of their ages.

(Present)   Let x be the son's age

Let 3x be Mrs. Brown's age

(Past)   Ten years ago

x - 10 - son's age

3x - 10 - Mrs. Brown' 5 age

Relationship: Mrs. Brown = 5 (son's age)

$3x - 10 = 5(x - 10)$ Substitution

$3x - 10 = 5x - 50$

$3x - 10 + 10 = 5x - 50 + 10$

$3x = 5x - 40$

$3x - 5x = 5x - 5x - 40$

$-2x = -40$

$$\frac{-2x}{-2} = \frac{-40}{-2}$$

$x = 20$ (son's age)

$3x = 3(20) = 60$ (Mrs. Brown's age)

Example #3:

The sum of 4 times Ken's age and 3 times Joyce's age is 80. Ken is 1 year less than Joyce's age. Find their ages.

(Present)   Let x be Joyce's age

Let x - 1 be Ken's age

Relationship: 4(Ken's age) + 3(Joyce's age) = 80

$$4(x - 1) + 3(x) = 80$$

$$4x - 4 + 3x = 80$$

$$7x - 4 = 80$$

$$7x - 4 + 4 = 80 + 4$$

$$7x = 84$$

$$\frac{7x}{7} = \frac{84}{7}$$

x = 12 (Joyce's age)

x - 1 = 12 - 1 = 11   (Ken's age)

NOTE: Always remember to assign the first variable to the one you know the least about.

Example #4:   The sum of 5 times Bob's age and 2 times Craig's age is 124. Bob is 1 year older than Craig. Find their ages.

(Present)   Let x be Craig's age

Let x + 1 be Bob's age

Relationship: 5 (Bob's age) + 2 (Craig's age) = 124
$$5(x + 1) + 2(x) = 124$$

$$5x + 5 + 2x = 124$$

$$7x + 5 = 124$$

$$7x + 5 - 5 = 124 - 5$$

$$7x = 119$$

$$\frac{7x}{7} = \frac{119}{7}$$

x = 17   (Craig's age)

x + 1 = 17 + 1 = 18   (Bob's age)

Let's Practice #49

1) Jim is five years older than Bill. In two years their ages will be 51. How old is each now?

2) Peter is seven years younger than Mike. Five years ago their ages totaled 25. How old is each now?

3) Marcus is half as old as his father. Next year, their ages will total 62. How old is each now?

4) Gene is ten years older than Rick. In six years, Gene will be twice as old as Rick. How old is each now?

5) Kimberly is 12 and her father is 37. How long will it be before her father is 2 times as old as she is?

6) Lisa is five years older than Annie and three years younger than Tina. Seven years ago, Tina's age totaled the ages of Lisa and Annie. How old is each now?

7) The sum of 4 times Cindy's age and 3 times Jill's age is 47. Jill is 1 year less than twice as old as Cindy. Find each of their ages.

8) The sum of 6 times Larry's age and 5 times Chris' age is 63. Larry is 1 year less than 3 times as old as Chris. Find each of their ages.

9) The sum of 4 times Sheila's age and 7 times Kathy's age is 169. Kathy is 1 year more than twice as old as Sheila. Find each of their ages.

10) The sum of 3 times Carolyn's age and 7 times Peggy's age is 137. Carolyn is 2 years less than twice as old as Peggy. Find each of their ages.

Coin Problems

In order to solve coin problems, we must set them up just as we do any other word problem.

Example #1:

Mr. Jones puts $5.75 in quarters in a pay telephone. How many quarters was this?

Let n be the number of quarters

Since quarters are represented with $.25 the relationship is: $n(.25) = \$5.75$

Now we have:	.25n	=	$5.75
Remember to move the decimal:	.25n	=	$5.75
	.25		.25

```
        23 .
     ┌────────
.25. │ 5 .75 .
        5 0
       ─────
          75
          75
         ───
           0
```

n = 23 quarters (Answer)

Example #2:

A cash box contained $12.25 in quarters, dimes and nickels. If there were five more than twice as many dimes as nickels and one less than three times as many quarters as nickels, how many of each kind of coin was there?

What do we know the least about? We know the least about the nickels.

Let n be the nickels

Let 2n + 5 be the dimes

Let 3n - 1 be the quarters

Relationship: Nickels + dimes + quarters = $12.25

$$.05n + .10(2n + 5) + .25(3n - 1) = 12.25$$

$$.05n + .20n + .50 + .75n - .25 = 12.25$$

$$1.00n + .25 = 12.25$$

NOTE: 1.00n = 1n = n

Add	
+ .05n	+ .50
+ .20n	- .25
+ .75n	+ .25
+1.00n	

So now we have:

$$n + .25 - 12.25$$

$$n + .25 - .25 = 12.25 - .25$$

$$n = 12.00$$

THIS IS NOT $12.00

Remember to always look back at the Let statements:

$$n = 12 \text{ nickels}$$

$$2n + 5 = 2(12) + 5$$

$$= 24 + 5$$

$$= 29 \ \text{dimes}$$

$$3n - 1 = 3(12) - 1$$

$$= 36 \quad - 1$$

$$= 35 \quad \text{quarters}$$

12 nickels, 29 dimes, and 35 quarters is the answer.

You need to read back through this example several times. Take your time and go through it carefully. Pay close attention to the Let statements. Remember to assign the first variable to the one you know the least about.

Let's Practice #50:

1) Mr Smith put $9.25 in quarters in a parking meter during the period of one month. How many quarters was this?

2) Mark had three less than twice as many nickels as dimes. If the total value of his coins was $1.45, how many of each kind of coin did he have?

3) Peggy Sue had 7 more than twice as many quarters as dimes. If the total value of her coins was $10.15, how many of each kind of coin did she have?

4) Mr. Williams put $7.25 in quarters in a parking meter during the period of one month. How many quarters was this?

5) Tina had three times as many nickels as dimes. If the total value of her coins was $1.00, how many of each kind of coin did she have?

6) Mike had five times as many quarters as dimes. If the total value of his coins was $16.20, how many of each kind of coin did he have?

7) A piggybank contained $14.55 in quarters, dimes, and nickels. If there were three more than twice as many dimes as nickels and three less than three times as many quarters as nickels, how many of each kind of coin was there?

8) Ebony's piggybank contained $8.80 in quarters, dimes, and nickels. There were two more than five times as many nickels as quarters and four less than twice as many dimes as quarters. How many of each kind of coin was there in the bank?

MANIPULATING FORMULAS

In order to manipulate formulas, we must recall our lessons on solving equations.

Example #1:   Solve:

$$y + 2b = b \text{ for } y$$

Subtract 2b from both sides of the equation: $y + 2b - 2b = b - 2b$

$$y = b - 2b$$

Answer: $y = -b$

Work like this can be very helpful in Physics and Chemistry.

Let's look at some more examples:

Example #2:   Solve: $s = \dfrac{1}{2} \, at^2 \text{ for } a$

$\left(\text{Remember} - \dfrac{1}{2} x = \dfrac{x}{2}\right):$ $s = \left(\dfrac{1}{2} t^2\right) a$

So $\dfrac{1}{2}t^2 = \dfrac{t^2}{2}$: $s = \left(\dfrac{t^2}{2}\right) a$

$$\dfrac{2}{t^2} \left(\dfrac{s}{1}\right) = \left(\dfrac{\cancel{2}}{\cancel{t^2}}\right)\dfrac{\cancel{t^2}}{\cancel{2}} a$$

NOTE: $2 \div 2 = 1$
$t^2 \div t^2 = 1$
leaving a

Answer: $\dfrac{2s}{t^2} = a$ **OR** $a = \dfrac{2s}{t^2}$

140

Example #3: Solve:

$$\frac{x+d}{3} = c \text{ for } x$$

1. Clear the denominator: $\left(\dfrac{\overset{1}{\cancel{3}}}{1}\right) \dfrac{x+d}{\underset{1}{\cancel{3}}} = \dfrac{c}{1} \left(\dfrac{3}{1}\right)$

$$x + d = 3c$$

2. Subtract d from both sides: $x + d - d = 3c - d$

3. Answer: $x = 3c - d$

Let's Practice #51:

Solve for the variable in bold print:

1) $x + c = a;$ **x**
2) $y - 2b = b;$ **y**

3) $2a - z = a;$ **z**
4) $a - y = 3y;$ **y**

5) $\dfrac{a + z - b}{2} = c;$ **z**
6) $I = Prt;$ **t**

7) $v = \dfrac{1}{3} bh;$ **h**
8) $v^2 = u^2 + 2as;$ **s**

9) $a = p + prt;$ **t**
10) $s = (n - 2)180;$ **n**

11) $c = 2\pi r;$ **r**
12) $v = \dfrac{1}{3}\pi r^2 h;$ **h**

13) $t = 2\pi r^2 + 2\pi rh;$ **h**
14) $p = \dfrac{3}{4} r + 25;$ **r**

This completes our study of high school algebra, at this time. Be sure to reread the Algebra Tutor carefully. Keep it as a handy reference to the many mathematical problems you will face all during your schooling and well into the particular job-world that you select for yourself. Mathematics, in the form of algebra, is one of the most valuable tools with which you can arm yourself to adequately prepare for your future. Algebra, in one form or another, is an integral part of almost every position available in the job market today. With the progress of modern technology, it has become more and more mandatory to have a thorough understanding of algebra. Your time will be well spent with the study of mathematics. As the year's roll by, you will be able to use the formulas you have learned, and practiced in these lessons, all through your life. Don't forget to review this book as often as needed.

GOOD LUCK!
WILLIE L. THOMAS

Let's Practice #1:   (Page 2)

1)	6	2)	13	3)	35	4)	44
5)	18	6)	17	7)	5	8)	4
9)	36	10)	8	11)	1008	12)	$\frac{4}{9}$

Let's Practice #2:   (Pages 6)

1)   8ab, 7ab, 5ba; -9a, -3a; -4b

2) $2r^2s^2$, $-6r^2s^2$, $-4r^2$, $2r^2$, $-2r^2$, $4r^2s$, $3r^2s$; $3s^2$

3)   a - 2b

4)   35x - y   OR   35x + -y

5)   -5p - q   OR   -5p + -q

6)   -2b

NOTE: Writing two signs can allow more room for mistakes. Let's do this.
When we see two signs coming together such as 2x + -3

OR 2x + (-3), let's multiply the two signs and get a one sign answer. Thus: (+) x (-) = -, so 2x + -3
OR 2x + (-3) can be written as 2x - 3. This is the most standard way to write the answer.

7)	-2h - 8m	8)	11a - 2b	9)	$-3a^2 - 21d$

10)   75 - 93rp

Let's Practice #3: (Page 9)

1) $6x - 1$
2) $4x + 6$
3) $5x - y$

4) $3a + 2b$
5) $5a - b + 1$
6) $10x + y + 5$

7) $8a^2 + 2a + 1$
8) $2x^2 y^2 + z^2$
9) $2x^2$

10) $4a^2 + 2b^2$
11) $3a^2 + 3b - 3$
12) $4x - 3$

Let's Practice #4: (Pages 11)

1) $4x + 3$
2) $2x + 2$
3) $3x + 3y$

4) $a + 4b$
5) $a - 3b + 5$
6) $-4x + 3y + 5$

7) $2a^2 - 2a + 5$
8) $-3z^2$
9) $4y^2 - 2z$

10) $-8ab$
11) $a^2 - b - 5$
12) $2x + 4y + 1$

Let's Practice #5: (Pages 12)

1) a^{10}
2) b^7
3) $24x^6$

4) $-a^4 b^6$ (Remember: $- \cdot - \cdot - = + \cdot - = -$)
5) $20a^3$

6) $2x^7 y^4$
7) $a^6 b^4$
8) a^{2n}

9) a^{3n}
10) x^{4n+4}
11) y^{6n}

12) $x^{2n} + 1$

Let's Practice #6: (Page 15)

1) $x^2 + 8x + 15$
2) $x^2 + 4x - 21$
3) $x^2 + 2x - 3$

4) $a^2 - 7a + 12$
5) $a^2 + a - 30$
6) $x^2 - 5x - 14$

7) $6x^2 - 7x - 3$
8) $20a^2 + 14a - 12$
9) $6c^2 + c - 12$

10) $x^3 + 5x^2 + 5x - 2$
11) $a^3 - 5a^2 + 7a - 3$

12) $x^4 - 3x^3 - 17x^2 + 12x - 2$

Let's Practice #7:   (Page 17)

1) $4a$

2) $3a$

3) $-3x^3$

4) x^3

5) $-\dfrac{1}{b^3}$

6) $\dfrac{a^3}{b^4}$

7) $-\dfrac{8r^2}{7t^4}$

8) $\dfrac{1}{3y^2}$

9) $\dfrac{a^3 b^5}{2}$ OR $\dfrac{1}{2}a^3 b^5$

10) $-\dfrac{4u^4}{3v^2}$

11) $\dfrac{2}{3b^3}$

12) $\dfrac{xz^2}{2y^2}$

Let's Practice #8:   (Page 18)

1) $\dfrac{2}{a^3}$

2) $a^2 b^3$

3) $3x^4 y^5$

4) $\dfrac{x}{5y^3}$

5) $\dfrac{6}{a^7}$

6) $\dfrac{15}{a^5 b^8}$

7) $\dfrac{1}{x^4 y^8}$

8) a^2

9) b

10) $\dfrac{x^3}{y^3}$

11) $\dfrac{1}{xy^4}$

12) $\dfrac{6x^3}{y^7}$

Let's Practice #9:   (Pages 19)

1) $2a + 4$

2) $-6a^2 + 4a + 1$

3) $2a + 3$

4) $3a^2 - a$

5) $x + 2$

6) $y^2 + x$

7) $4a^2 - 2a + 1$

8) $4x - 3x^2 + 2x^3$

9) $x^3 - 3x^2 y^2 + y^3$

10) $x + 2x^2 y^2 + 3y^2$

11) $-3x^2 y^2 + 6xy - 2x^2$

12) $2x^2 - 6y^2 - 3x^4$

Let Practice #10: (Pages 22)

1) $x + 1$

2) $x + 3$

3) $x - 9$

4) $y - 9$

5) $y + 3$

6) $y - 1$

7) $a+1$

8) $b + 4$

9) $2a + 2 + \dfrac{5}{3a - 1}$

10) $c + 2$

11) $x - 3 + \dfrac{1}{2x + 1}$

12) $m + 5$

Let's Practice #11: (Page 23)

1) $x = 2$

2) $t = -1$

3) $a = -3$

4) $z = 4$

5) $x = -7$

6) $b = -24$

7) $d = 46$

8) $b = +24$

9) $k = -11$

10) $y = -13$

11) $t = -12$

12) $t = 15$

Let's Practice #12: (Page 23)

1) $x = 6$

2) $z = 12$

3) $t = 15$

4) $w = 6$

5) $x = 105$

6) $x = 108$

7) $a = 18$

8) $t = 48$

9) $t = 28$

10) $x = 12$

11) $x = 16$

12) $t = 25$

Let's Practice # 13: (Pages 25)

1) $x = 3$

2) $x = 21$

3) $y = 3$

4) $x = 5$

5) $x = 6$

6) $x = 3$

7) $a = -18$

8) $y = 8$

9) $x = 3$

10) $c = 15$

11) $x = -4$

12) $y = -9$

Let's Practice #14:   (Page 26)

1) $x = 6$

2) $y = 3$

3) $z = -8$

4) $m = -7$

5) $n = -1$

6) $a = 15$

7) $a = 4$

8) $y = 5$

9) $x = 0$

10) $a = -3$

11) $b = 8$

12) $a = -2$

Let's Practice #15:   (Page 29)

1) $y = -7$

2) $x = -1$

3) $x = 4$

4) $a = -3$

5) $a = 7$

6) $a = 8$

7) $y = 2$

8) $y = -7$

9) $y = 2$

10) $x = -24$

11) $x = 0$

12) $y = -2$

Let's Practice 16:   (Page 31)

1) $6a(1 + 2a)$

2) $3b(5b - 3)$

3) $5a(a + 2)$

4) $x(x - 4y)$

5) $3h(hk + 1)$

6) $2(x + 3)$

7) $2x(x + 5y)$

8) $3ab(4a + 5b)$

9) $3y^2(2y^2 + 3y - 4)$

10) $5t^2(7 + 5t - 3t^2)$

11) $ax(a^2x^2 + ax - 2)$

12) $2(2x^2 + 2x - 3)$

Let's Practice #17:   (Page 33)

1) $(a - 1)(a + 1)$

2) $(x-4)(x + 4)$

3) $(y + 3)(y - 3)$

4) $(b^2 - 7)(b^2 + 7)$

5) $(5x + 3)(5x - 3)$

6) $(6a - 4)(6a + 4)$

7) $(4a + b)(4a - b)$

8) $(r^3 - t^3)(^3 + t^3)$

9) $(2x - y^2)(2x + y^2)$

10) $(7m^2 + n^3)(7m^2 - n^3)$

11) $(a^n + 6)(a^n - 6)$

12) $(x^{2n} - 3y^n)(x^{2n} + 3y^n)$

Let's Practice #18:   (Page 35)

1) $(x + 1)(x + 6)$ 2) $(x + 4)(x + 2)$

3) $(y - 2)(y - 1)$ 4) $(a + 4)(a - 2)$

S) $(b + 2)(b - 3)$ 6) $(c - 8)(c + 3)$

7) $(y + 4)(y + 5)$ 8) $(y + 5)(y + 2)$

9) $(y + 9)(y - 3)$ 10) $(d + 2)(d - 16)$

11) $(a + b)(a + b)$ 12) $(a - 7b)(a - 7b)$

Let's Practice #19:   (Page 38)

1) $(2a + 1)(a - 3)$ 2) $(a + 5)(3a - 2)$

3) $(2a + 1)(a + 3)$ 4) $(5x + 1)(x + 2)$

5) $(2x + 1)(2x - 5)$ 6) $(9x + 4)(2x - 3)$

7) $(4c + 3)(9c - 8)$ 8) $(6c + 7)(2c - 3)$

9) $(8c + 3)(3c - 5)$ 10) $(y - 8x)(y + 7x)$

11) $(1 - 4y)(6 + y)$ 12) $(3y + 2)(3y - 1)$

Let's Practice # 20:   (Page 39)

1) $(a + b - 3)(a + b + 2)$ 2) $(x - 7)(x + 7)$

3) $(c + d + 1)(c + d - 6)$ 4) $(x - 8)(x + 8)$

Let's Practice #21:   (Page 40)

1) $(4x + 4y + 3)(x + y - 2)$ 2) $(3x + 3y + 1)(x + y - 6)$

2) $(2a + 5)(a + 6)$ 4) $(5a - 2)(a + 2)$

Let's Practice #22:   (Page 41)

1) $2a + 1$ 2) $2x - 1$

3) $-1(2x + 5)$ OR $-(2x + 5)$ OR $-2x-5$

4) $(x+ 7)(5x - 1)$

Let's Practice #23:   (Page 44)

1) $(y + z)(2 +x)$ 2) $(a + c)(a + b)$

3) $(y^2 + 4)(2y+ 1)$ 4) $(3c - 4)(a - 2b)$

5) $(x + 2)(x -2)(2x - 1)$ 6) $(2a^2 + 5)(a + 3)$

Let's Practice #24:   (Page 45)

1) $(x + 2)(a + b)$ 2) $(3x + 2)(x + 1)$

3) $(x + y)(2y -1)$ 4) $(a - 1)(b + 3)$

5) $(1 + y)(x - 6)$ OR $(y+ 1)(x - 6)$

6) $(-a + b)(x - 2)$ OR $(b - a)(x - 2)$

Let's Practice #25:   (Page 47)

1) $2(2x - 1)^2$ 2) $(x + 1)(x - 1)(x + 2)(x - 2)$

3) $2(a + 2b - 2)(a - 2b + 2)$ 4) $(x + 3z - 2)(y - 3z)$

5) $(y - 2)(y + 2)(y - 1)$ 6) $2(a - 2)(a + 2)(a^2 + 1)$

Let's Practice #26: (Page 50)

1)   {6, -6}   2)   {7, 8}   3)   {0, 7}

4)   {0, 2}   5)   {$\frac{-1}{3}$, $\frac{3}{2}$}   6)   {0, $\frac{4}{3}$}

7)   {$\frac{-2}{3}$, 3}   8)   { 0, $\frac{5}{3}$, -2 }

Let's Practice #27: (Pages 53)

1)   {7, 9}   2)   {7, -9}   3)   {-3, 8}

4)   {$\frac{3}{2}$, $\frac{-3}{2}$}   5)   {2}   6)   {$\frac{1}{3}$, -1}

7)   {0, 1, -1}   8)   {0, 2, -2}   9)   {$\frac{-1}{2}$, 3}

10)   {0, 7 }   11)   {-3, 3, 2, -2}   12)   {2, -2, -1}

Let's Practice #28: (Pages 58)

1)   r = 10.0   2)   10.3   3)   26.0

4)   y = 4.0   5)   10.2   6)   r = 13.9

7)   12.0   8)   2.2   9) $\frac{1}{2}$

10) $\frac{4}{5}$

Let's Practice #29: (Page 62)

1) $3, (x \neq 2)$

2) $\frac{1}{3}, (x \neq -\frac{y}{3}, y \neq -3x)$

3) $\frac{x-y}{x+y}, (x \neq -y, y \neq -x)$

4) $\frac{x+3}{x-6}, (x \neq 0, x \neq 6)$

5) $\frac{y-x}{y+x}, (y \neq 0, x \neq -y, y \neq -x)$

6) $\frac{1}{3a-1}, (a \neq 0, a \neq \frac{1}{3})$

7) $\frac{1}{3a-4}, (a \neq -\frac{4}{3}, a \neq \frac{4}{3})$

8) $\frac{4}{y-5}, (y \neq -5, y \neq 5)$

9) $\frac{2(y+2)}{y+6}, (y \neq 2, y \neq -6)$

10) $\frac{x-5}{x+7}, (x \neq -3), x \neq -7)$

11) $\frac{4(4x-y)}{y-x}, (y \neq -4x, y \neq x)$

$(x \neq y, x \neq -\frac{y}{4})$

12) $1, (a \neq b, a \neq -b)$

Let's Practice #30: (Page 65)

1) $\frac{2x^3}{y^2}$

2) $\frac{y}{y-2}$

3) $\frac{2}{x+1}$

4) $\frac{a+b}{3}$

5) $\frac{2(x-y)}{x}$

6) x^2

7) $\frac{(a-b)(a^2+b^2)}{2(a+b)}$

8) $\frac{x(x+3)}{4}$

9) $\frac{x-3}{x-2}$

10) $\frac{-2(2y+1)}{y}$

11) $\frac{6(a-4)(a+2)}{(a-2)(a+1)}$

12) $\frac{y^2+3}{6y}$

Let's Practice #31: (Pages 68)

1) 2

2) $\dfrac{2}{a+1}$

3) $\dfrac{x+2}{2x}$

4) $(a-b)(a+b)$

5) $\dfrac{x+2}{x+3}$

6) $\dfrac{a^2}{6}$

7) $\dfrac{3y}{7}$

8) $2(a+b)$

9) 1

10) 1

11) $\dfrac{2x-3}{x}$

12) a

Let's Practice #32: (Page 72)

1) $\dfrac{a^2-2a}{(a-1)(a+1)}$

2) $\dfrac{2y+1}{y^2+y}$

3) $\dfrac{1}{6(x+1)}$

4) $\dfrac{7x-5}{(x+5)(x-5)}$

5) $\dfrac{x^2+2xy-y^2}{(x+y)(x-y)}$

6) $\dfrac{-a+2}{a(a-2)(a+1)}$

7) $\dfrac{y+x}{x^2-xy}$

8) $\dfrac{3x-1}{(x-1)(x+1)^2}$

9) $\dfrac{3a+5}{(a+1)(a-2)(a+2)}$

10) $\dfrac{8a-10}{a(a-5)(a+5)}$

11) $\dfrac{-x^2+7x+6}{(x+1)(x+2)^2}$

12) $\dfrac{y+1}{y-1}$

Let's Practice #33: (Pages 76)

1) $y=4$

2) $y=21$

3) $y=1$

4) $x=\dfrac{28}{15}$

5) $x=\dfrac{15}{32}$

6) $y=\dfrac{27}{2}$

7) $x=5$

8) $y=\dfrac{11}{10}$

9) $a=3$

10) $x=-5$

11) $x=\dfrac{5}{-21}$

12) $a=\dfrac{7}{5}$

Let's Practice #34:   (Pages 79)

1) $x = 15$ 2) $x = 16$ 3) $x = -\dfrac{15}{11}$

4) $a = -8$ 5) $a = -3$ 6) $x = \dfrac{1}{2}$

7) $x = \dfrac{1}{2}$ 8) $y = 12$ 9) $y = -\dfrac{2}{3}$

10) $y = 3$ 11) $a = 7$ 12) $a = -8$

Let's Practice #35:   (Page 87)

1) $a = 3$ 2) $y = 6$ 3) $x = -6$

4) $x = \dfrac{16}{5}$ 5) $a = \{-4, 1\}$ 6) $a = \{1, 4\}$

Let's Practice #36:   (Page 91)

Problems 1 - 10:

Let's Practice #36: (Page 91)

Problems 1 - 10:

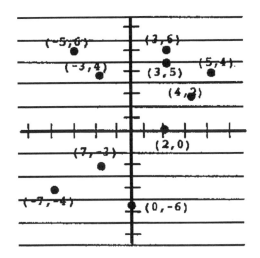

Let's Practice #37: (Page 94)

1)

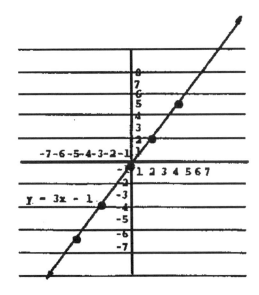

2)

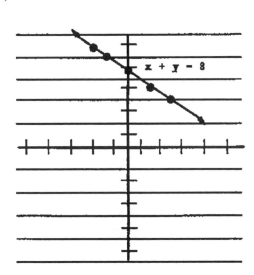

Let's Practice 37: (Continued from Page 94)

3)

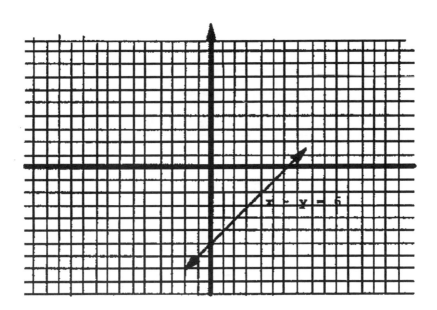

4)

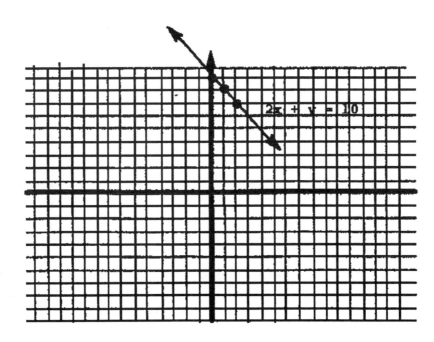

5)

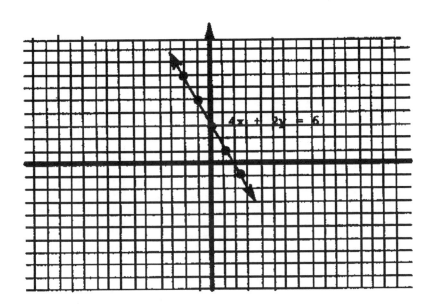

6)

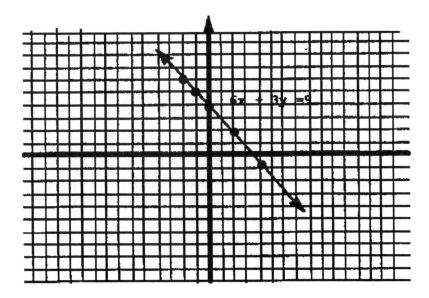

Let's Practice #38: (Page 98)

Problems: 1 - 6

1) Solution: (3, 1)

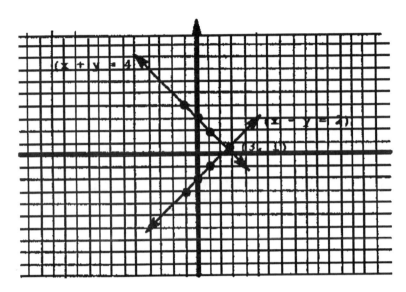

2) No Solution (Parallel line)

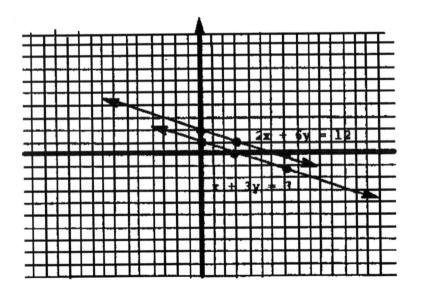

Let's Practice #38: (Continued from Page 98)

3)

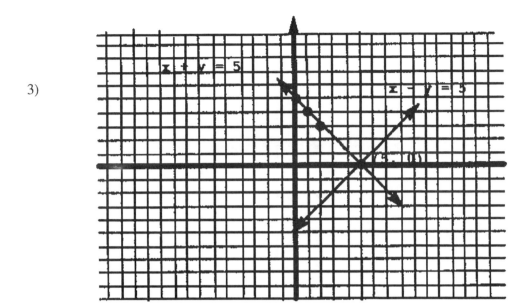

Solution: (5, 0)

4)

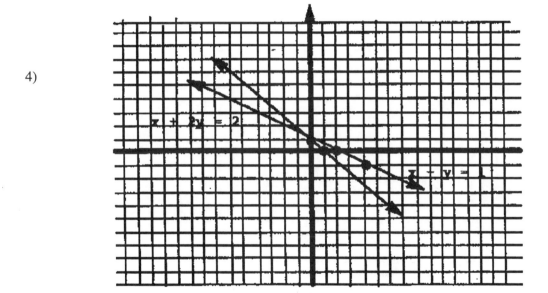

Solution: (0, 1)

Let's Practice #38:   (Continued from Page 98)

5)

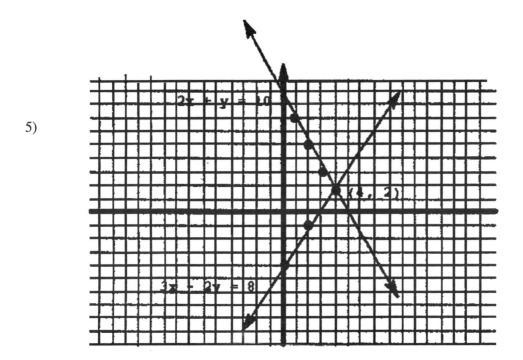

Solution: (4,2)

6)

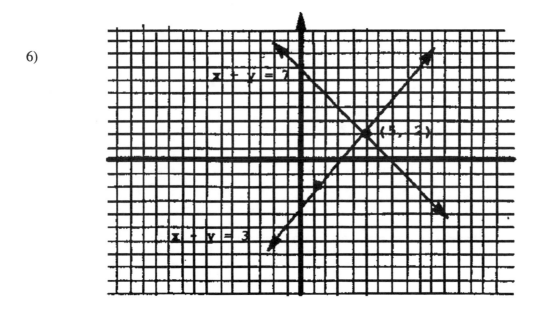

Solution: (5, 2)

Let's Practice #39:   (Page 100)

1)   (6, 3)

2)   (-3, -1)

3)   (4, 0)

4)   (-1, 2)

5) $\left(-1, \dfrac{2}{3}\right)$

6) $\left(\dfrac{5}{3}, \dfrac{2}{3}\right)$

7)   (6, 4)

8)   (0, 0)

9)   (-4, -3)

Let's Practice #40:   (Page 103)

1)   (6, -2)

2)   (0, -4)

3)   (2, 1)

4)   (2, 3)

5)   (6, 2)

6)   (2, 2)

7)   (-1, -1)

8) $\left(3, \dfrac{1}{3}\right)$

9)   (-2, -3)

10)   (3, 8)

11)   (4, -3)

12)   (0, -1)

Let's Practice #41:   (Page 106)

1)   (-3, 2)

2)   (5, 1)

3)   (-4, 0)

4)   (0, -2)

5)   (3, 2)

6)   (0, -3)

7)   (2, -2)

8)   (2, 0)

9)   (-1, -1)

10)   (-2, 1)

11)   (-6, 0)

12)   (1, -1)

Let's Practice #42:   (Page 110)

1)   0

2)   No slope

3)   -2

4) $-\dfrac{2}{3}$

5) $-\dfrac{1}{2}$

6) $\dfrac{7}{2}$

7)   4

8)   1

9)   2

10)   -2

11) $\dfrac{2}{3}$

12)   4

Let's Practice #43: (Page 114)

1) $15x - 3y = 2$

2) $2x + 5y = -20$

3) $y = -3$

4) $y = -2x - 4$

5) $y = \frac{3}{2} x + 3$

6) $y = \frac{1}{2} x + 3$

NOTE: $\frac{x}{2} = \frac{1}{2}x$

Let's Practice #44: (Page 118)

1) $3x - y = 3$

2) $x + 2y = -1$

3) $y = -2$

4) $5x - 4y = 25$

5) $x + 4y = 6$

6) $8y - 3y = 22$

Let's Practice #45: (Page 119)

1) $x < 8$

2) $x > 9$

3) $y > -4$

4) $x \leq 3$

5) $x < 0$

6) $x > -6$

7) $x \geq \frac{4}{3}$

8) $x \geq 6$

9) $x \geq -14$

10) $y \geq 13$

Let's Practice #46: (Pages 122)

1) $3n$

2) $n + 7$

3) $5n$

4) $n - 2$

5) $\frac{1}{2} n - 8$ **OR** $\frac{n}{2} - 8$

6) $n + 4$

7) $n + 6 = 50$

8) $n - 9 = 72$

9) $3n + 7 = 25$

10) $n - 5 = 12$

11) $2n - 8 = 20$

12) $5n - 1 = 14$

Let's Practicee: #47:   (Page 123)

1)   10   2)   8   3)   15

4)   18   5)   9   6)   21

Let's Practice: #48:   (Pages 129)

1)   17, 18   2)   15, 17, 19   3)   4, 6, 8, 10

4)   10, 11, 12   5)   22, 23, 24, 25   6)   5, 6, 7, 8

7)   -3, -1, 1, 3   8)   23 wide and 25 long

Let's Practice: #49:   (Pages 137)

1)   Bill is 21 and Jim is 26.

2)   Mike is 21 and Peter is 14.

3)   Marcus is 20 and his father is 40.

4)   Rick is 4 and Gene is 14.

5)   13 years.

6)   Annie is 10, Lisa is 15, and Tina is 18.

7)   Cindy is 5 years old and Jill is 9 years old.

8)   Larry is 8 years old and Chris is 3 years old.

9)   Sheila is 9 years old and Kathy is 19 years old.

10)   Carolyn is 20 years old and Peggy is 11 years old.

Let's Practice: #50:   (Pages 139)

1)   37 quarters

2)   8 dimes, 13 nickels

3)   14 dimes, 35 quarters

4)   29 quarters

5)   4 dimes, 12 nickels

6)   12 dimes, 60 quarters

7)   15 nickels, 33 dimes, 42 quarters

8)   13 quarters, 22 dimes, 67 nickels

Let's Practice: #51:   (Pages 141)

1) $x = a - c$

2) $y = 3b$

3) $z = a$

4) $y = \dfrac{a}{4}$

5) $z = 2c + b - a$

6) $t = \dfrac{1}{pr}$

7) $h = \dfrac{3v}{b}$

8) $s = \dfrac{v^2 - u^2}{2a}$

9) $t = \dfrac{a - i}{pr}$

10) $n = \dfrac{s + 360}{180}$

11) $r = \dfrac{c}{2\pi}$

12) $h = \dfrac{3v}{\pi r^2}$

13) $h = \dfrac{t - 2\pi r^2}{2\pi r}$

14) $r = \dfrac{4p - 100}{3}$

GLOSSARY

Abscissa:	The coordinate of the point on the horizontal axis where a vertical line, from a given point, meets the horizontal axis.
Absolute Value:	The positive number of any pair of opposite real numbers is called the absolute value of each of the numbers. For example: the absolute value of 1 **OR** -1 IS 1.
Angle:	A figure formed by two lines starting at the same point.
Axiom:	A statement that is assumed to be true.
Binomial:	A polynomial with two terms. For example: x+ 2, a + 3, y - 2, etc.
Consecutive even Integers:	Obtained by counting by two's from any even integer.
Consecutive Integer:	Obtained by counting by one's from any given integer.
Consecutive Odd Integers:	Obtained by counting by two's from any odd integer.
Constant:	A numerical expression without a variable. For example: 1, 5, 8, 12, $\frac{1}{2}$ -1, $-\frac{3}{4}$ etc.
Coordinate:	The number assigned to a point on the number line.
Coordinate Axis:	The axes (horizontal and vertical) of a coordinate system set up in a plane.
Coordinate Plane:	A plane in which a coordinate system has been set up.
Coordinate System:	A system of graphing ordered pairs of numbers in relation to two axes (horizontal and vertical) that intersect at right angles at their

Coordinate of a Point:	point of origin. The numbers where the coordinate meets, written as an ordered pair of numbers.
Degree of a Polynomial:	The greatest of the degrees of its terms after it has been simplified.
Denominator:	When the fraction $\overline{P}\atop q$ is written, q is the denominator where $q \neq 0$.
Difference:	For any two real numbers a and b, the difference a - b is the number whose sum with b is a.
Equal Expressions:	Expressions that name the same number.
Equation:	A statement formed by placing an equality symbol between two numerical or variable expressions.
Evaluate an Expression:	Replace each variable in a particular expression by the numeral for a given value of the variable, and simplify the results.
Factor:	When two or more numbers are multiplied, they each are factors of the product.
Factored Form:	The expression "a • a • a" is the factored form of the third power of a, i.e. a^3.
Factoring:	The act of breaking down the factors of a number or an expression over a specified set.
Formula:	An expression of numerical relationships between quantities such as physical or other measurements
Fractional:	An expression in the form $\overline{p}\atop q$ $q \neq o$

Fractional Equation:	An equation which has a variable in the denominator of one or more terms.
Graph of a Number:	The point on the number line paired with the number.
Graph of an Equation:	All points, and only those points, whose coordinates satisfy the equation.
Graph of an Ordered Pair of Numbers:	The point in the plane paired with an ordered pair of real numbers.
Greatest Common Factor:	The greatest integer which is a factor of each of two or more integers.
Greatest Monomial Factor of a Polynomial:	The monomial factor having the greatest numerical coefficient and the greatest degree.
Grouping Symbol:	A device used to enclose a numerical expression. Examples include parentheses, brackets, and fraction bars.
Hypotenuse:	The right side of a triangle opposite the right triangle.
Identity:	An equation which is true for every numerical replacement of the variable or variables.
Identity Element:	0 is the identity element for addition; 1 is the identity element for multiplication; For example: 1) $0 + a - a$ (addition) 2) $1 \cdot a = a$ (multiplication)
Inequality:	A statement formed by placing an inequality symbol between two numerical or variable expressions.
Integers:	The numbers found on the number line, such as: (-2, -1, 0, 1, 2

Inverse Operations:	Operations that "undo" each other, for example, addition and subtraction or multiplication and division.
Least Common Denominator: (L.C.D)	The least positive common multiple of the denominator of two or more fractions.
Linear Equation in One Variable:	A polynomial equation of degree one.
Linear Equation in Two Variable:	Any equation equivalent to one of the form ax + by - c, where a, b, and c are real numbers with a and b not both zero.
Linear Term:	A term of degree one in the variable:
Member of an Inequality:	The expressions joined by an inequality symbol.
Member s of an Equation:	The expressions joined by the symbol of equality.
Monomial:	A single term which is either a numeral, a variable, or a product of a numeral and one or more variables. Examples: 5a, x, 2x, $-3x^2$, etc.
Numeral:	A name for a number.
Numerator:	In the fraction $\frac{p}{q}$, q is the numerator.
Numerical Coefficient:	The non-variable factor in a term. In short, the number that sits in front of the variable, or on the left side of the variable. (It is understood to be 1, if there is not another number showing.) For example: In 3ab, the numerical coefficient is 3, in x, it is 1.
Open Sentence in Two Variables:	An equation of inequality which contains two variables.

Opposite of a Number:	The same number with an opposite sign: -3, +3, -5, +5.
Ordered Pair:	A pair of elements in which the order is specified.
Ordinate:	The coordinate of the point on the vertical axes where a horizontal line from a given point meets the vertical axis.
Origin:	The starting point, labeled "0" on a number line the zero point of both of two number lines that intersect at right angles.
Parallel Lines:	Lines that lie in the same plane, but have no point in common. (Do not cross)
Perimeter:	The perimeter of a geometric figure is the distance around it.
Plotting a Point:	Locating the graph of an ordered pair of real numbers on a coordinate plane.
Polynomial:	A sum of monomials.
Power:	A product in which all the factors, except 1 are the same. Such as: the fifth power of 6 is defined by $6^5 = 6 \cdot 6 \cdot 6 \cdot 6 \cdot 6$.
Pythagorean Theorem:	In any right triangle, the square of the hypotenuse equals the sum of the squares of the other two. sides.
Quadratic Polynomial:	A polynomial of degree two in the variable.
Quadratic Term:	A term of degree two in the variable.
Quotient:	The quotient of $a \div b$, $b \neq 0$, is the number whose product with b is a.
Real Number:	Any number paired with a point on the number line.

Relation:	Any set of ordered pairs. The set of the first components is called the "domain" and the set of second components is called the if "range".
Replacement Set:	The set of numbers that the variable may represent.
Satisfy:	Each component in the solution set of an open sentence satisfies that sentence.
Second Component:	The second number in an ordered pair.
Similar Terms:	Terms that are exactly alike, meaning the variable parts must be the same, but their numerical coefficient can be different.
Simplify:	Replace a numerical expression by the most common name of its value.
Solve:	Find the solution set of an open sentence over a given domain.
Square Root:	The number x is the square root of the number y if $x^2 = y$.
Substitution Method:	A method for finding the solution of a pair of linear equations in two variables. Solve one pair of linear equations, then substitute the resulting expressions in the remaining set of linear equations, solve the resulting equation, and finally finding the corresponding value of the other variable.
Term:	A mathematical expression using numerals or variables or both to indicate a product or a quotient.
The Product of a Constant and Variable:	An expression which gives the result of a constant being multiplied by a variable. Example: "3 • n" is usually written as '3n'. This is the product of the constant 3 and the variable n.
Theorem:	A statement that is proven to be true using axioms, definitions, and other theorems in a logical development.

Transformation:	Producing an equation equivalent to the original equation: By adding the same number to each member; by dividing each member by the same nonzero number, by multiplying each member by the same nonzero number, by substituting either member by an expression equal to it, or by subtracting the same number from each member.

Trinomial: A polynomial with three terms: $x^2 + 2x + 4$, $3x^2 - 6x + 9$, etc.

Variable: A symbol used to represent; one or more unknown numbers. Examples: x, y, a, b, c, etc.

X-Intercept: The abscissa of the point of intersection of a graph with the x-axis.

Y- Intercept: The ordinate of the point where a line crosses the y-axis.